What Every Engineer
Should Know About

Finite Element Analysis

WHAT EVERY ENGINEER SHOULD KNOW
A Series

Editor

William H. Middendorf

Department of Electrical and Computer Engineering
University of Cincinnati
Cincinnati, Ohio

Other volumes in preparation

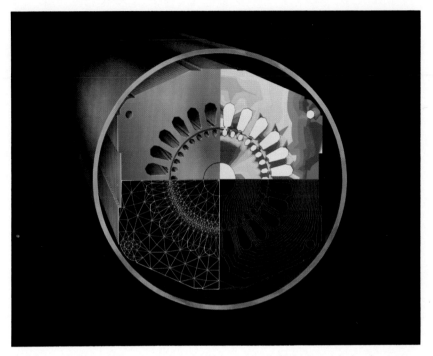

Plate 1   Finite element analysis of the magnetic field in an induction motor (see pages 164–167).

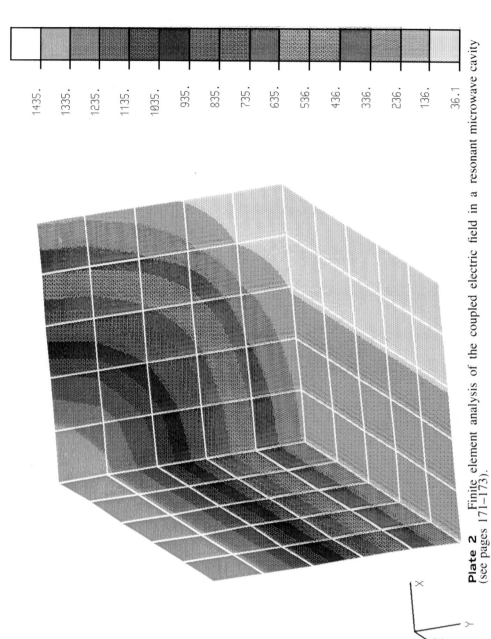

Plate 2 Finite element analysis of the coupled electric field in a resonant microwave cavity (see pages 171–173).

What Every Engineer Should Know About

Finite Element Analysis

edited by

John R. Brauer

The MacNeal-Schwendler Corporation
Milwaukee, Wisconsin

MARCEL DEKKER, Inc. New York and Basel

ISBN 0-8247-7832-4

MARCEL DEKKER, INC.
270 Madison Avenue, New York, New York 10016

Current printing (last digit):
10 9 8 7 6 5 4

PRINTED IN THE UNITED STATES OF AMERICA

Preface

This book is intended to serve the large variety of engineers, managers, designers, and students who need to use finite elements to analyze engineering problems. No prior knowledge of finite element analysis is required to use this book. Only basic engineering knowledge is assumed.

Unlike many books that present in great detail the mathematical theory behind finite element computer programs, this book emphasizes how such programs are applied. Many examples covering a wide variety of engineering disciplines are presented to illustrate how finite element analysis is used as an engineering design tool.

After two introductory chapters explaining the background and key concepts of the finite element method, the bulk of the book is devoted to use of finite element analysis in various branches of engineering. Each chapter on a particular engineering application area covers the types of problems analyzed and how to perform an analysis, illustrated by many examples. Final chapters summarize present and future aspects of finite element software, computer hardware, and managing their use.

The balanced emphasis on applications for structural, thermal, electrical, and fluid engineers is aimed at showing the multidisciplinary nature of finite element analysis. The design of industrial products for today's competitive world market requires the diversified finite element analysis techniques described in this book.

The editor and contributors wish to thank numerous colleagues for their help and cooperation in developing the techniques described here. In particular, we thank Dr. R. Y. Bodine for his many years of support.

John R. Brauer

Contents

Contributors

G. E. Barron

John R. Brauer

Nancy J. Lambert

Vern D. Overbye

Glenn H. Stalker

The MacNeal-Schwendler Corporation
Milwaukee, Wisconsin

What Every Engineer
Should Know About

Finite Element Analysis

1 History and Development

G. E. Barron

The MacNeal-Schwendler Corporation, Milwaukee, Wisconsin

What every engineer should know about finite element analysis begins with its history. Despite its currently visible success in analyzing difficult engineering problems in structural mechanics, heat transfer, electromagnetics, fluid flow, and other areas, it really hasn't existed very long. Chapter 8 will discuss more on the possible future contributions of the method, now popularly called the finite element method, sometimes referred to as FEM or FEA for finite element analysis, but for now let's focus on its history and the motivation that led to the analytic tool as we know it today. In addition to What, every good reader learns that Who, When, and Why are also critical to understanding a story, and we will look at all of these. Especially we will look at the Whos, because they made it happen and also made the period of time when this was unfolding very exciting.*

WHO AND WHEN

Today textbooks on the finite element method abound, and most engineering schools offer courses on the subject, yet 25 years ago this was not the case.

---

*The contributor of this chapter began engineering school in 1959 at the University of Kansas and completed his M.S. in Aerospace Engineering in 1966 under Dr. James Tiedemann, thesis on finite elements, and because of learning about finite element analysis as it was emerging would like to share the personal excitement and knowledge of some of the early pioneers.

Similarly, many commercial computer software products exist today that implement the finite element approach to solving engineering problems. Any computer codes in the late 1950s and early 1960s for solving structural analysis problems in a manner similar to what is now called the finite element approach existed only within the aircraft industry, which was one of the earliest motivated to have improved analysis methods.

The reference credited by many with having really established the method is the Turner, Clough, Martin, and Topp paper [1], published in 1956. The aircraft industry's influence may be seen here in that Turner worked for Boeing Aircraft and Martin was at the University of Washington Aeronautical Engineering Department in Seattle. Professor Ray Clough, at the University of California-Berkeley's Civil Engineering Department, was not in the aerospace camp* and yet was a key Who in the development of the method. His 1960 paper [2] coined the term "finite element method." No one was certain at the time that this term would remain; however, it has, and it makes Clough's contribution seem even more monumental.

Other Whos include Argyris and colleagues from the University of Stuttgart in West Germany. His book with Kelsey [3] was published in 1960, but was based on work published in 1954 and 1955 in the *Aircraft Engineering Journal*. This work, certainly contemporary with Ref. [1], deserves much credit even though the name finite element was coined elsewhere. Reference [4] demonstrates how close-knit these early pioneers were and again how strong an influence the aircraft industry exerted. This reference was the proceedings of a conference held by the Air Force. Two hundred scientists from several countries attended and Turner, Clough, and Argyris were there. By this time developments of substantial magnitude were beginning to be made to implement the finite element method for use on computers, including the NASA project that led to NASTRAN (NASA STRuctural ANalysis, proprietary to NASA).

A commercial software product, ASKA (automatic system for kinematic analysis, but really a finite element software system), which is still supported and used today came from Argyris' work. The well-known program called SAP (structural analysis program) came from Clough's student Ed Wilson at University of California-Berkeley.

Professor R. D. Cook at the University of Wisconsin wrote a textbook [5] for teaching finite elements in college in 1974 after the method was well established. This is an excellent additional source of information on finite

---

*One of Prof. Clough's graduate students, J. L. Tocher, worked for Boeing after Berkeley, so the ties must have been close.

elements. His historical review suggests that the idea really goes back to a mathematician, R. Courant, in a paper [6] in 1943. Courant proposed breaking a continuum problem into triangular regions and replacing the fields with piecewise approximations within the triangles.

In summary, then, the inception of the technique or method is 30–40 years old, at least, and it was probably established by several pioneers almost independently. Possibly nothing was done with Courant's ideas for a decade because computers large and fast enough were not available until the 1950s and then only available in large aircraft companies. This gives some answers on Who and When, but let's move on to the What and Why.

WHAT

What finite elements are was hinted at in Courant's idea. Typically engineering problems of mechanics, solid or fluid, have been addressed in the past by deriving differential equations relating the variables of interest after using reasoning that appeals to physics and engineering principles. The principles used to establish valid equations describing the behavior of the engineering problem at hand include equilibrium, Newton's idea regarding force acting on a mass, potential energy, strain energy, conservation of total energy, virtual work, thermodynamics, conservation of mass, Maxwell's equations, and many more. The problem seemed to always be that once all the hard work of formulating the problem was complete, solving the resulting mathematical equations, usually differential equations (sometimes linear and often nonlinear partial differential equations), was almost impossible. Only very simple problems of regular geometry (rectangular, circular, etc.) with the most simple boundary conditions could be solved.

The solid foundation in structural mechanics leading up to the development of finite elements was actually done by pioneers in the last part of the nineteenth century. The excellent Dover Publications [7] has made Castigliano's gigantic contributions to potential and complementary energy principles and the theory of equilibrium available for our review. This reference also reviews the work of a dozen or more other workers during this time period. Castigliano's principles were published originally in Italy over 100 years ago, in 1879. He solved many practical problems in railway bridges and trusses using his theories. Reference [8] indicates that the Lord Rayleigh style of obtaining real answers, perhaps more clearly than Castigliano's, was to assume a shape of, say, the lateral displacement of a column or shaft and then obtain answers through minimization of the energy computed using this function. Great skill was required to select the function so that boundary conditions were also satisfied. The dilemma grew worse when the geometry was other than a straight rod. The first paragraph of Chapter VII of Ref. [8] indicates that Rayleigh or his followers realized that the geometrical shape of many members made them difficult or impossible to handle with an appropriate untheoretical approximation.

Appendix II of Ref. [8] notes that Ritz in 1909 made an important extension to Rayleigh's principles which included using multiple independent functions allowing more than one frequency of a shaft to be computed. The disadvantage noted was the need to solve an increasingly large number of simultaneous algebraic equations. Ritz could have been the true father of finite elements if only he had extended his idea to use different geometric regions, establish separate approximating functions in each region, and then hook them together. This is precisely what finite elements are about. They are building blocks, as will be described in Chapter 2, for creating the overall complex geomety of the engineering problem at hand, much in the same way that children build windmills, trucks, and so forth from toy blocks and sticks. The idea, which Courant apparently got right 40 years later than Ritz, had to wait until modern digital computers took away the fear of large numbers of algebraic equations. Soon matrices and matrix methods of organizing large numbers of algebraic equations were brought into the finite element approach; recall that the word "matrix" was in the title of the important Air Force Conference in 1965. Now mathematicians reentered the scene to remind the engineers of all the methods of solving matrix equations from linear algebra. An excellent reference for this important branch of mathematics is Ref. [9], a book translated from the Russian author V. N. Faddeeva.

WHY

Restated, the finite element method is one wherein the difficulty of mathematically solving large complex geometric problems (say doing the stress analysis of a Boeing 747) is transformed from a differential equation approach to an algebraic problem, wherein the building blocks or finite elements have all the complex equations solved for their simple shape (say a triangle, rod, beam, etc.). The representation of the relationship of the important variables for the little, but not infinitesimal, elements is determined through a Rayleigh or Ritz approach just for each element. Once this is done, a matrix of size equal to the number of unknowns for the element can be produced which represents the element. This is now a linear algebraic relation and not a differential equation. The entire problem can thus be cast as a larger algebraic equation by assembling the element matrices within the computer in much the same way that the real problem is built with many simple pieces of material. Chapter 2 goes through this with an example. The key here is the simplicity and elegance of the approach. Specialists such as modern-day Castiglianos can get the finite elements formulated correctly, and computers can be taught to assemble and solve the many thousands of equations represented by the overall problem. The engineer of today can concentrate on the objective of the analysis, specifying the real-world loads and boundary conditions, and then obtain answers that normally would have required a prototype. Indeed, modern graphic portraits of the model inside the computer make one believe it is real.

The Why seems clear now as to the motivation for developing the finite element method. The aircraft industry simply could not wait until the first Boeing 747 was ready for testing to validate its structural integrity. The advantage of doing the analysis early* is that it is fairly simple to change loads, materials, and geometry and recompute stresses for a modified airplane. The advantages remain today, and the method has spread to other industries and other applications than just structures.

In fact, the method can be used to solve almost any problem that can be formulated as a field problem. The development of additional software products and use within industry has taken the last 25 or 30 years. There has to be some critical need to motivate its use. The nuclear industry certainly needed it. The product ANSYS (see Chapter 3) grew out of the nuclear industry. The automotive industry began using the approach in earnest in the early 1970s when the combination of quality Japanese imports, fuel concerns due to the oil embargo, as well as safety dictated a better approach than build and test.

The area still under the early phase of use is probably the electromechanical and magnetic area, probably owing to delayed recognition of need and transfer of knowledge from structural to electrical engineers. Today's increased world competition, use of rare earth magnets, the pressure to design for electromagnetic compatibility, and new electronic devices such as high-frequency power supplies are changing this trend.

The use of finite elements for flow analysis outside of the large aerospace firms is similarly in an embryonic stage. This development of more general

---

*In the World War II effort to build aircraft, the stress analysis was often not complete until long after the planes were flying, and many approximations were made, say, for cutouts for doors in an otherwise cylindrical fuselage.

analysis of flow probably had to wait for very fast computers such as the CRAY, since most problems are nonlinear.

Indeed, finite element analysis would not be where it is today if computers had not proliferated and become faster and less expensive beyond belief.

REFERENCES

1. Turner, M. J., R. W. Clough, H. C. Martin, and L. J. Topp, "Stiffness and Deflection Analysis of Complex Structures," *Journal of the Aeronautical Sciences,* Vol. 23, No. 9, September 1956.
2. Clough, R. W., "The Finite Element Method in Plane Stress Analysis," *Proceedings of the Second Conference on Electronic Computation,* ASCE, 1960.
3. Argyris, J. H., and S. Kelsey, *Energy Theorems in Structural Analysis,* Butterworth, London, 1960.
4. *Proceedings of the Matrix Method in Structural Mechanics Conference,* AFFDL-TR-66-80, held at Wright-Patterson Air Force Base, Ohio, 1966.
5. Cook, R. D., *Concepts and Applications of Finite Element Analysis,* John Wiley & Sons, Inc., New York, 1974.
6. Courant, R., "Variational Method for the Solution of Problems of Equilibrium and Vibrations," *Bulletin of the American Mathematical Society,* Vol. 49, 1943.
7. Castigliano, C. A., *Theory of Equilibrium Equations of Elastic Systems and Its Application,* Dover Publications, Inc., New York, 1966 (a translation by E. S. Andrews from the Italian book of the same title published by Frederic Negro in Turin, Italy, 1879, was published also in London, 1919, by Scott, Greenwood, and Son).
8. Temple, G., and W. Bickley, *Rayleigh's Principle.* Dover Publications, Inc., New York, 1956 (a new edition based on original work in 1933).
9. Faddeeva, V. N., *Computational Methods of Linear Algebra.* Dover Publications, Inc., New York, 1959.

2 Basic Finite Element Concepts

John R. Brauer

The MacNeal-Schwendler Corporation, Milwaukee, Wisconsin

NEED FOR ANALYSIS OF FIELDS

Engineering design is aided by engineering analysis, the calculation of performance of a trial design. To predict the performance it is often necessary to calculate a field, which is defined as a quantity that varies with position within the device analyzed.

There are many kinds of fields, and each field has a different influence on the device performance. Table 2.1 lists the various fields which will be analyzed in this book. The mechanical stresses of the device being designed should be calculated to make sure the device will not break. The heat flow should be calculated to make certain that the device will not be too hot. In electrical and magnetic devices the electromagnetic fields should be calculated. In fluid devices the fluid flow should be calculated.

Closely associated with any field is a potential. Table 2.1 also lists the potentials commonly associated with each of the above fields. The fields are related to the potentials as their derivatives with respect to position. The exact form of the spatial derivative may vary with the type of problem, as will be discussed in later sections.

Table 2.1   Various Aspects of Performance

Field	Potential
Heat flux	Temperature
Mechanical stress	Displacement
Electric field	Voltage
Magnetic field	Magnetic vector potential
Fluid velocity	Fluid potential

FINITE ELEMENT MODELLING

Calculation of all the above fields and potentials can be performed using finite element analysis. The analysis begins by making a finite element model of the device. The model is an assemblage of finite elements, which are pieces of various sizes and shapes. The finite element model contains the following information about the device to be analyzed:

geometry, subdivided into finite elements
materials
excitations
constraints

Material properties, excitations, and constraints can often be expressed quickly and easily, but geometry is usually difficult to describe.

Figure 2.1 shows a typical engineering problem that happens to be a static thermal or heat transfer problem. To perform finite element thermal analysis, the finite element model shown in Figure 2.2 was constructed by dividing the device into finite elements. The finite elements can be very small where small geometric details exist and can be much larger elsewhere. In each finite element a simple variation of potential (temperature) is assumed, as described on page 13. The corners of the finite elements are called grid points or nodes. The task of the finite element computer program is to solve for all the unknown grid point potentials and finite element fields.

The finite elements in Figure 2.2 are two dimensional, consisting of triangles and quadrilaterals, as described on page 12. One-dimensional finite elements are described on page 11, and three-dimensional finite elements are described on page 15. Each finite element has a material property, which may be different or the same as the properties of the other finite elements.

Excitations may be present in the finite element model. They may occur either within finite elements or at grid points. For example, in the thermal

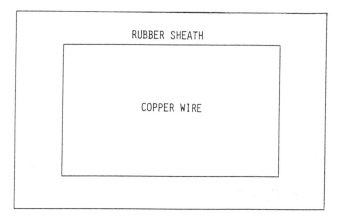

Figure 2.1   Cross-section of current-carrying copper wire with a rubber cover sheath. The wire and its current extend into and out of the page.

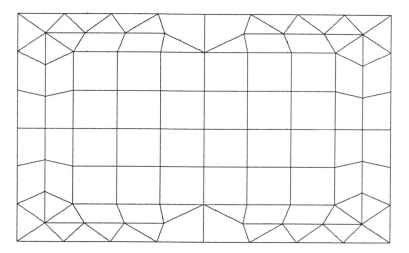

Figure 2.2   Two-dimensional finite element model of Figure 2.1.

problem of Figures 2.1 and 2.2 the finite elements made of copper wire have heat generation that excites the temperatures. Excitations can also be called loads, particularly in mechanical stress analysis.

Each grid point in a finite element model has a potential with one or more degrees of freedom. In thermal problems the potential is temperature, a scalar with one degree of freedom per grid point. In structural problems the potential is a displacement vector with up to 6 degrees of freedom per grid point. The 6 degrees of freedom are translations in three directions and rotations about the three directions.

Each degree of freedom at a grid point may be unconstrained (unknown) or constrained. One type of constraint is a single-point constraint (SPC), which is a known value of a degree of freedom. An SPC in a thermal problem is a known temperature at a grid point. Another type of constraint is a multipoint constraint (MPC), in which a degree of freedom (component) of a potential is unknown except that it is equal to or a function of another potential component, usually at another grid point. Usually most grid points in a model are unconstrained, and the finite element program will determine the values of all their degrees of freedom.

ENERGY FUNCTIONAL MINIMIZATION

The way finite element analysis obtains the temperatures, stresses, fields, or other desired unknown parameters in the finite element model is by minimizing an energy functional. An energy functional consists of all the energies associated with the particular finite element model.

The law of conservation of energy is a well-known principle of physics. It states that, unless atomic energy is involved, the total energy of a device or system must be zero. Thus, the finite element energy functional must equal zero.

The finite element method obtains the correct solution for any finite element model by minimizing the energy functional. Thus, the solution obtained satisfies the law of conservation of energy.

The minimum of the functional is found by setting the derivative of the functional with respect to the unknown grid point potential to zero. It is known from calculus that the minimum of any function has a slope or derivative equal to zero. Thus, the basic equation for finite element analysis is

$$\frac{\partial F}{\partial p} = 0 \tag{2.1}$$

where F is the functional and p is the unknown grid point potential to be calculated. The above simple equation is the basis for finite element analysis.

The functional F and unknown p vary with the type of problem, as will be described in the next sections.

In variational calculus [1] the functional is shown to obey a relationship called Euler's equation. Substitution of the functional in the appropriate Euler's equation yields the differential equation of the physical system. Thus, the finite element solution obeys the appropriate differential equation.

ONE-DIMENSIONAL SPRING ELEMENT

Figure 2.3 shows a simple mechanical spring. It is an example of a one-dimensional finite element, which is one-dimensional because it connects only two grid points, here labeled grid points a and b.

The unknown grid point potential in a mechanical or structural problem such as the spring is displacement (positional change). In Figure 2.3 the displacement can only be in the x direction; that is, the displacement is also in one dimension only and thus has only 1 degree of freedom.

The energy functional is always the total energy. Here the functional F is made up of the energy stored in the spring and the energy input in compressing or expanding the spring:

$$F = \frac{1}{2} kx^{**}2 - \int f dx \tag{2.2}$$

where f is force and k is the spring stiffness. Substituting Eq. (2.2) in Eq. (2.1) where $p = x$ gives

$$kx - f = 0 \tag{2.3}$$

which gives the familiar expression for spring displacement:

$$x = \frac{f}{k} \tag{2.4}$$

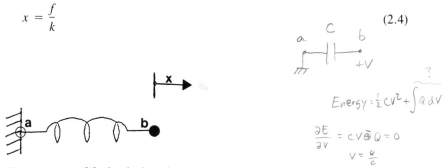

Figure 2.3 Mechanical spring connecting two points, acting as a one-dimensional finite element.

Thus, we have seen in this simple one-dimensional example of a spring how an energy functional obtains a correct solution for the unknown grid point potential, which is here the displacement component x.

TWO-DIMENSIONAL POISSON'S EQUATION ELEMENT

Two-dimensional finite elements were previously shown in Figure 2.2. By definition, two-dimensional finite elements connect three or more grid points lying in a two-dimensional plane. Here we will briefly derive the equations for a triangular finite element modeling a physical system that obeys Poisson's differential equation.

Poisson's equation in two dimensions is

$$\frac{\partial}{\partial x} k \frac{\partial T}{\partial x} + \frac{\partial}{\partial y} k \frac{\partial T}{\partial y} = -P \qquad (2.5)$$

where x and y are the two dimensions and k is a material property. Equation (2.5) governs static temperatures T, in which case P is power input per unit volume. It also governs static electric or magnetic fields, in which case T is potential and P is charge density or current density, respectively.

The energy functional for all three kinds of physical problems is

$$F = \int_s \left[\left(\frac{1}{2} \right) k (\text{del } T)^{**}2 \right] dS - \int_s \left(\frac{1}{2} PT \right) dS \qquad (2.6)$$

where S is the surface area in the two dimensions. The first term of Eq. (2.6) is the energy stored in the cases of electric or magnetic fields and is related to power dissipated in the case of thermal fields. The second term is the input energy in the cases of electric or magnetic fields and is related to power input in the case of thermal fields. The first term involves the gradient

$$\overline{G} = \text{del } T = \nabla T \qquad (2.7)$$

$$\overline{G}(x,y) = \frac{\partial T}{\partial x} \overline{u}_x + \frac{\partial T}{\partial y} \overline{u}_y \qquad (2.8)$$

where $\overline{u}_x$ and $\overline{u}_y$ are unit vectors.

The simplest type of two-dimensional finite element assumes a linear, or first-order, variation of the unknown potential T over the element. Figure 2.4 shows a triangular finite element. It may be of any size and shape, except that for accurate results each of its three angles should be much less than

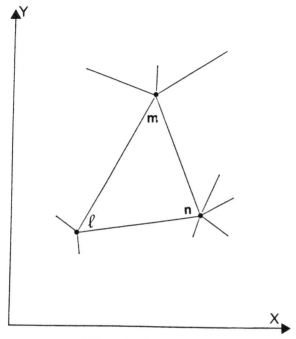

Figure 2.4 Triangular finite element in the xy plane.

180°. Within this first-order element T is related to the three unknown T values at the three triangular grid points according to

$$T = \sum_{k=\ell,m,n} T_k (a_k + b_k x + c_k y) \tag{2.9}$$

Evaluating Eq. (2.9) at the three vertices gives the solution for the a, b, c coefficients:

$$\begin{bmatrix} a_\ell & a_m & a_n \\ b_\ell & b_m & b_n \\ c_\ell & c_m & c_n \end{bmatrix} = \begin{bmatrix} 1 & x_\ell & y_\ell \\ 1 & x_m & y_m \\ 1 & x_n & y_n \end{bmatrix}^{-1} \tag{2.10}$$

Substituting Eq. (2.9) in Eq. (2.7) gives

$$\overline{G}(x,y) = \sum_{k=\ell,m,n} [b_k T_k \overline{u}_x + c_k T_k \overline{u}_y] \tag{2.11}$$

Thus, the temperature gradient is constant within a particular triangular finite element. Quadrilateral finite elements are composed of two or four triangles.

The grid point potentials T_k can be found by minimizing the functional (2.6), where $ds = dxdy$. Substituting Eq. (2.11) and Eq. (2.6) in Eq. (2.1) and considering one triangular finite element yields

$$\int_s \frac{\partial}{\partial T_j} \left[\frac{kG^2}{2} - \frac{PT}{2} \right] ds = 0, \qquad j = \ell, m, n \tag{2.12}$$

Carrying out the integration over the triangle can be shown to yield the 3-by-3 matrix equation [2]

$$[S]\,[T] = [P] \tag{2.13}$$

where the "stiffness" matrix is

$$[S] = k\Delta \begin{bmatrix} (b_\ell b_\ell + c_\ell c_\ell) & (b_\ell b_m + c_\ell c_m) & (b_\ell b_n + c_\ell c_n) \\ (b_m b_\ell + c_m c_\ell) & (b_m b_m + c_m c_m) & (b_m b_n + c_m c_n) \\ (b_n b_\ell + c_n c_\ell) & (b_n b_m + c_n c_m) & (b_n b_n + c_n c_n) \end{bmatrix} \tag{2.14}$$

where Δ is the area of the triangle, and the right-hand side is the "load vector":

$$[P] = \frac{\Delta}{3} \begin{bmatrix} P \\ P \\ P \end{bmatrix} \tag{2.15}$$

Equation (2.13) solves for the potential T in a region containing the one triangle with modes ℓ, m and n in Figure 2.4. For practical problems with N nodes (grid points), the above process is repeated for each finite element, obtaining a stiffness matrix $[S]$ with N rows and N columns. $[P]$ and $[T]$ are then column vectors containing N rows. Note that Eq. (2.13) is similar to Eq. (2.3). In both cases "stiffness" times an unknown potential equals a forcing function.

AXISYMMETRIC FINITE ELEMENTS

All devices designed by engineers are in reality three-dimensional. A special case of a three-dimensional device is one that has axial symmetry.

An axisymmetric finite element is a fairly simple extension of the two-dimensional triangular element described in the previous section. A typical axisymmetrical finite element is shown in Figure 2.5. The axis of symmetry is the y axis, and the element is assumed to revolve 360° around that axis. Thus, devices made up of cylinders, cones, or other shapes all with the same axis of symmetry can be modeled with axisymmetrical finite elements. The

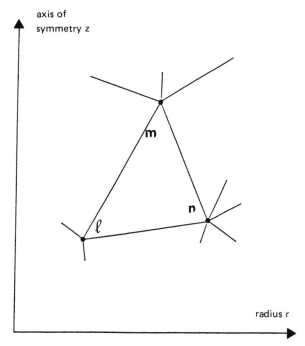

Figure 2.5 Axisymmetric finite element.

derivation of the axisymmetric finite element matrix equation is similar to that of two-dimensional elements, except all areas become volumes equal to the area times 2π times the element radius [3].

THREE-DIMENSIONAL FINITE ELEMENTS

A three-dimensional finite element has at least four grid points, and they do not all lie in one plane. The element forms a solid shape containing a volume of material.

The one-dimensional, two-dimensional, and axisymmetric finite elements described above are used whenever possible, because they are simpler to describe and use than three-dimensional finite elements. Three-dimensional elements are unnecessary when the unknown grid point quantities are essentially invariant in at least one dimension. For example, devices with dimensions into the z direction much greater than their maximum x or y dimensions can usually be assumed to be two-dimensional, provided their excitations and constraints are also two-dimensional.

Of course, many devices to be designed by engineers are highly three-dimensional, where there is no axis of symmetry. In such cases truly three-dimensional finite elements must often be used.

Figure 2.6 shows the three most common three-dimensional solid finite elements. The simplest is the tetrahedron, or four-sided element with four grid points. It is similar to a pyramid, except that all four of its sides are triangles. Also shown is the pentahedron, or five-sided element with six grid points. Two of its sides are triangles and the other three sides are quarilaterals. The third solid element shown is the hexahedron, or six-sided element. This "brick" element has eight grid points, and all six of its sides are quadrilaterals. The sizes and shapes of these three-dimensional elements are arbitrary, except accuracy may be reduced if the triangular sides of a pentahedron or the opposite sides of a hexahedron are not parallel within 30° or so.

The derivation of the matrix equation for the tetrahedral finite element is rather similar to that of the triangular element. Thus, in the case of quantities obeying Poisson's equation the tetrahedral element is similar to the two-dimensional triangular element, except of course the spatial integration must be carried out in all three directions. Extending Eq. (2.9) gives

$$T = \sum_{k=\ell,m,n,o} T_k(a_k + b_k x + c_k y + d_k z) \tag{2.16}$$

The derivation of the matrix equations proceeds in a fashion similar to that described on page 14.

SOLUTION FOR ENTIRE MODEL

As shown in Figure 2.2, a finite element model consists of a large number of finite elements. The assumed simple variation of the unknown field within each finite element gives the potentials and fields less freedom than in the real device, and thus the calculated potentials and fields are usually slightly ~~less~~ than ~~in~~ reality. The accuracy of the calculation generally increases as the

different

Figure 2.6 Three-dimensional finite elements: the tetrahedron, the pentahedron, and the hexahedron.

number of elements in the model increases. For best accuracy the density of finite elements should be highest in regions where the potential gradient is highest.

As shown in the previous sections, all finite elements obey a matrix equation. The number of unknowns in the matrix equation for one element equals the number of grid points in the element times the number of degrees of freedom per grid point.

The matrix equation for the finite element model is an assembly of the matrix equations of all its finite elements. The total number of unknowns N can therefore run into hundreds or thousands or more.

Fortunately the N by N "stiffness" matrix is inherently sparse; that is, it contains many zero matrix elements. Only the nonzero matrix elements need be stored and manipulated by the computer. Often the computer program rearranges or "resequences" the matrix rows and columns to move the nonzero entries within a relatively narrow band along the matrix diagonal [4]. The width in columns of the resequenced matrix is called the resequenced bandwidth.

The computer time required to solve the matrix equation depends on the type of resequencing, the solution method (such as Gaussian elimination), the computer hardware, and other factors which will be mentioned in Chapter 7. An approximate relation, however, is that solution time is roughly proportional to the cube of the number of unknowns N, or at least to the square of N. Today even personal computers can easily solve problems with N equal to 500–1000 in a few hours. If N can be reduced by taking advantage of any geometric symmetry to solve only a fraction of the geometry by means of appropriate constraints, then substantial computer time and analyst time can be saved.

SOLUTION TYPES

The type of solution described in the above sections is a steady-state static solution. Several other types of solutions are obtainable by the finite element method.

The other solution types are for time-varying quantities. One type is called transient, in which the results are obtained by stepping along in time. An example would be the thermal problem of Figures 2.1 and 2.2 for input excitation power zero at time zero and then following a given function of time, with the resulting temperature distribution at all times to be determined. Another solution type is for excitations that are a sinusoidal function of time, as is common in electrical problems. In both types of time variation the solution is obtained by solving or resolving a real or complex matrix equation in a fashion similar to that described in the previous section.

Another type of problem—sometimes called an eigenvalue problem—is to determine natural or resonant frequencies of a physical system. The eigenvalues of the finite element matrix equation are determined by special techniques to obtain the natural frequencies and mode shapes.

Solution types depend on the physical system analyzed. The following chapters describe solutions for a wide range of engineering problems. Finite element analyses of structural, thermal, electromagnetic, and fluid problems are each described in separate chapters.

REFERENCES

1. Hildebrand, F. H., *Methods of Applied Mathematics,* Prentice-Hall, Inc., Englewood Cliffs, NJ, 1965, Chapter 2.
2. Huston, R. L., and C. E. Passerello, *Finite Element Methods—An Introduction,* Marcel Dekker, Inc., New York, NY, 1984, Chapter 5.
3. Brauer, J. R., "Finite Element Analysis of Electromagnetic Induction in Transformers," IEEE Winter Power Meeting, paper A77-122-5, February 1977.
4. Cuthill, E. H., and J. M. McKee, "Reducing the Bandwidth of Sparse Symmetric Matrices," Proc. of 24th National Conference of Association of Computing Machinery, 1969, pp. 157–172.

3 Structural Analysis

Vern D. Overbye

The MacNeal-Schwendler Corporation, Milwaukee, Wisconsin

STIFFNESS METHOD INTRODUCTION

The relationship between free end displacement and applied equilibrium force for a constrained spring was shown in Figure 2.3 and Eq. (2.4) in Chapter 2. This relationship is the basis of the stiffness method of structural analysis. It will be expanded by using a plane truss to introduce mathematical concepts needed for structural finite element analysis. It should be noted that the stiffness k is the slope of the spring force versus displacement curve.

Plane Truss Member Element and Global Stiffness

Many finite element texts [1–3] introduce the method by means of a truss. Historically, matrix methods of structural analysis were first applied to trusses, because truss members (or rods) are simple two-force elements joined by pins and loaded axially at the pinned ends.

Figure 3.1a shows a uniform, linearly elastic rod in a horizontal orientation with a force and displacement at each end (positive quantities are directed to the right). If the truss member has zero rigid body motion and is in force equilibrium, elementary mechanics texts show that a positive extension u from a positive member force p are related by

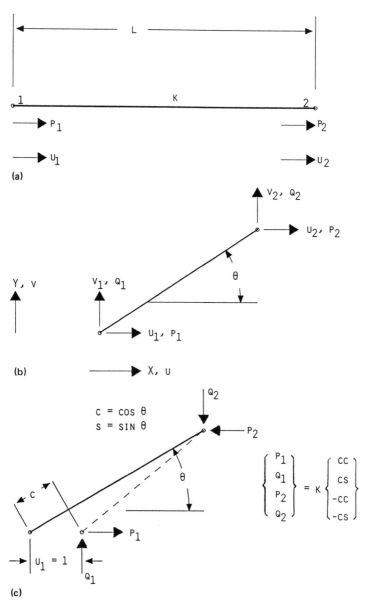

Figure 3.1 Planar truss member and global stiffness. (a) Horizontal orientation. (b) Inclined orientation. (c) Force equilibrium, enforced u_1.

$$p_1 = -p_2 = -p \tag{3.1}$$

$$u = u_2 - u_1 = \frac{pL}{AE} = \frac{p}{k}$$

where A, L, E, and k are rod cross-section area, length, material elastic modulus, and spring rate, respectively.

The two relationships expressed in Eq. (3.1), namely, $p_1 = ku_1 - ku_2$ and $p_2 = -ku_1 + ku_2$, may be expressed in matrix notation as

$$[k]\{u\} = \{p\} \tag{3.2}$$

where

$$[k] = k \begin{vmatrix} 1 & -1 \\ -1 & 1 \end{vmatrix}, \qquad \{u\} = \begin{Bmatrix} u_1 \\ u_2 \end{Bmatrix}, \qquad \{p\} = \begin{Bmatrix} p_1 \\ p_2 \end{Bmatrix} \tag{3.3}$$

The stiffness matrix $[k]$ has two rows and two columns because each end of the rod has 1 translational degree of freedom (DOF). The matrix is symmetrical and diagonal terms are positive. The first column in the matrix represents force equilibrium if u_1 is displaced one unit (u_2 constrained), and the second column represents force equilibrium if u_2 is displaced one unit (u_1 constrained). Positive diagonal terms indicate that a positive displacement requires a positive force.

The stiffness matrix of the Figure 3.1a rod may be called a local or element stiffness matrix [1], since it is valid along an axis passing through the two rod end grid points. This axis would have to be redefined along the rod axis if it is located at an arbitrary orientation in space.

A global stiffness matrix may be expressed along three orthogonal axes at each of the spatially located grid points. Each element attached to the grid point contributes stiffness. This technique will be illustrated using an inclined planar truss member and extending it to a three-rod planar truss.

Figure 3.1b shows a rod inclined at an angle θ in a Cartesian coordinate system XY plane. The notation shows u and v, and p and q, to represent displacement and force in the X and Y directions, respectively. This is a 4-DOF model, with each rod grid point having 2 translational DOF.

Figure 3.1c shows force equilibrium on the rod with u_1 displaced one unit. This results in a rod axial compressive force of $k \cos \theta$ (where it is assumed that θ has an insignificant change due to unit displacements). This axial force then has p_1, q_1, p_2, and q_2 equilibrium components at rod ends, as shown on the figure. Successive application of unit displacement at v_1, u_2 and v_2 (with the other 3 DOF constrained) results in a four row by four column stiffness matrix for the inclined rod.

$$[k] = k \begin{bmatrix} cc & cs & -cc & -cs \\ cs & ss & -cs & -ss \\ -cc & -cs & cc & cs \\ -cs & -ss & cs & ss \end{bmatrix} \tag{3.4}$$

where $c = \cos \theta$ and $s = \sin \theta$.

As expected, each column of $[k]$ sums to zero, the matrix is symmetrical, and diagonal terms are positive. Letting $\theta = 0$, Eq. (3.4) reduces to $[k]$ in Eq. (3.3), with a row and column of zeros for each of the two 0 DOF normal to the horizontal rod in Figure 3.1a. This insertion of rows and columns of zeros is referred to as matrix expansion [1].

The global stiffness matrix of the single planar rod as given in Eq. (3.4) will now be used as a means of constructing the stiffness matrix of a three-rod truss. Each rod stiffness matrix will be expanded to full model global size using rows and columns of zeros to account for DOF not affecting the particular rod being evaluated. This process is called the assembly procedure [1] or expansion to structure size [2].

Figure 3.2a shows the physical arrangement of a gantry that may be used on a mining dragline or construction crane. The structure consists of a heavy section compression member with attachment lugs to support boom cables. A lighter section tension member is attached to the top of the compression member, and both are supported on a base frame.

Figure 3.2b shows a planar statically determinant three-member truss to represent the gantry. A global stiffness matrix will be assembled for this truss using the inclined rod stiffness matrix given by Eq. (3.4) and assuming no supports.

Figure 3.3a shows successive integers used to label each of the three rods and grid points. The angle θ is given to describe inclination of each rod to the X axis. Each joint has a u (X-directed) and v (Y-directed) displacement; hence the complete displacement vector will consist of six components (u_1, v_1, u_2, v_2, u_3, and v_3). As before, each column in the final stiffness matrix will represent force equilibrium at a grid point DOF when it is displaced a unit distance. However, frame member columns, with linear and rotational DOF, do not sum to zero [2].

Figure 3.3b shows the element stiffness matrix for rod 1 in the truss using the appropriate θ from Figure 3.3a in Eq. (3.4). Note that rod 1 involves only u displacements at grid points 1 and 2 with matrix expansion of zeros for the other DOF such that the matrix is global size (six rows and six columns).

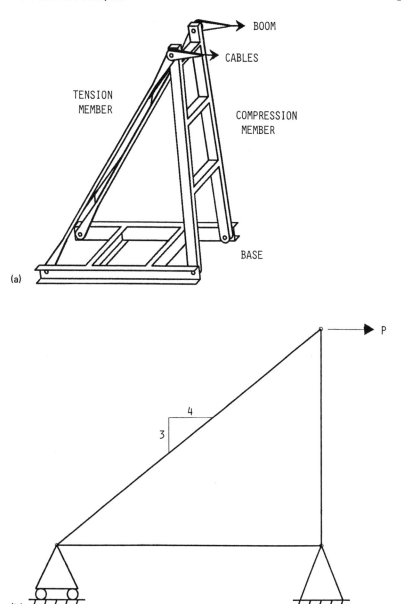

Figure 3.2 Construction equipment gantry and model. (a) Physical arrangement. (b) Statically determinant truss representation.

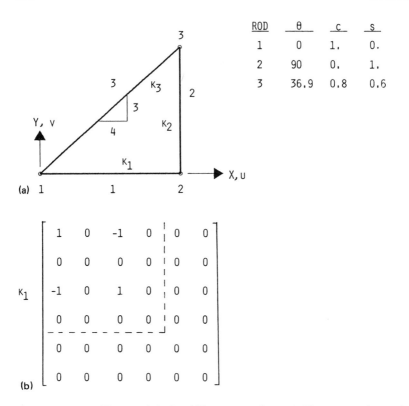

Figure 3.3 Truss global stiffness matrix. (a) Element orientation. (b) Element 1 stiffness matrix.

The global stiffness matrix of the entire truss will be a six row and six column matrix with each entry the sum of the contribution from each rod.
    Rod 2 yields

$$
k_2 \begin{bmatrix}
0 & 0 & 0 & 0 & 0 & 0 \\
0 & 0 & 0 & 0 & 0 & 0 \\
0 & 0 & 0 & 0 & 0 & 0 \\
0 & 0 & 0 & 1 & 0 & -1 \\
0 & 0 & 0 & 0 & 0 & 0 \\
0 & 0 & 0 & -1 & 0 & 1
\end{bmatrix}
\tag{3.5a}
$$

Rod 3 yields

$$
k_3
\begin{bmatrix}
(.8)(.8) & (.6)(.8) & 0 & 0 & -(.8)(.8) & -(.6)(.8) \\
(.6)(.8) & (.6)(.6) & 0 & 0 & -(.6)(.8) & -(.6)(.6) \\
0 & 0 & 0 & 0 & 0 & 0 \\
0 & 0 & 0 & 0 & 0 & 0 \\
-(.8)(.8) & -(.6)(.8) & 0 & 0 & (.8)(.8) & (.6)(.8) \\
-(.6)(.8) & -(.6)(.6) & 0 & 0 & (.6)(.8) & (.6)(.6)
\end{bmatrix}
\tag{3.5b}
$$

Finally, a six row by six column stiffness matrix $[K]$, which is the sum of all three rod elements, relates displacement and applied loads as

$$
[K]\{u\} = \{P\}
\tag{3.5c}
$$

Chapter Preview

This brief introduction to the finite element stiffness matrix and assemblage of individual elements into a global stiffness matrix for the complete structure will be elaborated in succeeding sections. Structural or stress analysis problem types that may be solved using commercial finite element programs will be discussed. Available elements, structure support techniques, solution techniques, and output data control are described. Logical techniques to evaluate the goodness of a solution are also outlined. Finally, a series of static and dynamic structural problems will be formulated, solved, and presented using a popular commercial finite element program.

PROBLEM TYPES

General Structural Analysis Equation

Finite element structural analysis (as introduced on p. 19) is a mathematical procedure to determine displacements, velocity, acceleration, stress, and support reaction forces due to applied mechanical and/or thermal loads that may or may not vary with time. The structure may be as simple as the single degree of freedom (SDOF) spring structure discussed in Chapter 2 (p. 11) or the 6-DOF truss described on page 22, or as complicated as a several hundred thousand multiple degree of freedom (MDOF) offshore oil drilling platform.

Equations of equilibrium may be derived for each DOF in the structure and expressed as a general structural analysis equation in matrix form:

$$[M] \{\ddot{u}\} + [B] \{\dot{u}\} + [K] \{u\} = \{P\} + \{N\} \tag{3.6}$$

where $[M]$, $[B]$, and $[K]$ are mass, damping, and stiffness matrices of the structure, respectively. Also, $[K] = [K_E] + [K_G]$, where $[K_E]$ is elastic stiffness and $[K_G]$ is geometric or differential stiffness (see p. 27). The symbols $\{\ddot{u}\}$, $\{\dot{u}\}$, and $\{u\}$ represent acceleration, velocity, and displacement vectors, respectively, for the N DOF in the structure. Right-hand side terms $\{P\}$ and $\{N\}$ refer to static and/or time-varying applied linear and nonlinear mechanical and/or thermal loads in Eq. (3.6). Constraint forces from structure supports are obtained in data recovery.

Various structural analysis problems are merely subsets of the general structural analysis equation and will be discussed in detail below.

Linear Statics

Equation (3.6) reduces for linear static analysis (introduced on p. 19) to

$$[K] \{u\} = \{P\} \tag{3.7}$$

when structural inertia effects are negligible, nonlinear loads are absent, and geometric stiffness effects may be ignored. Linear static analysis assumes that stress is proportional to strain (i.e., the materials follow Hooke's law) for the entire load history. Also, small displacements are assumed such that structure geometric deformation does not influence the vector of applied loads (zero follower forces) and additional internal moments do not result for initially colinear forces (geometric or differential stiffness, p. 27).

Inertia Relief

Inertia relief [3] is a subset of linear static analysis wherein the structure's internal inertia load distribution from rigid body acceleration balances applied loads. An example of inertia relief analysis would be a horizontal dragline boom given an angular acceleration by means of a moment applied about the vertical rotational axis at the support base. Structural deformation of the boom due to this angular acceleration is obtained from a static analysis rather than a more involved transient analysis.

Elastic Buckling

As mentioned above, the stiffness matrix consists of both elastic and geometric stiffness terms. The elastic stiffness terms depend only on material properties, whereas the geometric stiffness terms depend on the element's geometry, displacement, and state of stress [2]. Buckling failure of thin-wall structures often shows large bending deformation required to store released membrane energy.

Figure 3.4 demonstrates a simple geometric stiffness concept [4]. The figure shows a torsional spring on a massless pendulum with a downward-directed (positive) force. Displacement of the applied load is also positive downward. Equating the derivative of the energy functional to zero (see

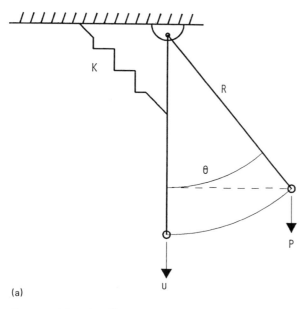

(a)

R = pendulum length
K = torsional spring constant
P = applied load
U = displacement at applied load

$F = \dfrac{K\theta^2}{2} - PR(\cos\theta - 1)$ energy functional [Eq. (2.2), Chapter 2]

$\dfrac{dF}{d\theta} = K\theta + PR\sin\theta = 0 \qquad \sin\theta \cong \theta$

$(K + PR)\,\theta = (K_E + K_G)\,\theta = 0$

where $K_E = K$ = elastic stiffness

$\qquad K_G = PR$ = geometric stiffness

$\qquad P_{CR} = -\dfrac{K}{R}$

(b)

Figure 3.4 Geometric stiffness example (a) Physical arrangement. (b) Geometric stiffness.

Chapter 2) results in a stiffness term K_E dependent on torsional spring material properties and a geometric stiffness term K_G dependent on applied load and geometry. A positive value for P stiffens the system, and a negative value for P reduces the stiffness. A nonzero $\delta\theta$ satisfies the relationship in Figure 3.4 if $P_{cr} = -\dfrac{K}{R}$ (which defines critical buckling load).

For elastic buckling the general structural analysis equation becomes

$$[K_E + \lambda K_G]\{u\} = 0 \qquad (3.8)$$

where λ is the applied load magnification factor. Note that it is assumed that the geometric stiffness is proportional to the applied load.

Equation (3.8) resembles real eigenvalue structural analysis (see next section) and its solution technique is quite similar. The smallest value obtained for λ determines the first buckling mode for the structure. The compression member in Figure 3.2a would present a buckling analysis problem.

Real Eigenvalues

Let all applied loads, geometric stiffness, and structural damping be zero. Further assume that displacement at all locations on the structure are sinusoidal time variations with the same frequency and phase relationship:

$$\{u\} = \{\phi\} \cos \omega t \qquad (3.9)$$

where $\{\phi\}$ is a vector of real numbers called an eigenvector. The general structural analysis Eq. (3.6) becomes

$$[K - \omega^2 M]\{\phi\} = 0 \qquad (3.10)$$

where ω is circular frequency.

Equation (3.10), which is similar to Eq. (3.8), is called an eigenvalue equation with the eigenvalue defined as $\lambda = \omega^2$. Roots of the matrix in Eq. (3.10) yield structural natural frequencies (see p. 37). Many numerical eigenvalue extraction techniques are available [5,6] to determine the discrete set of eigen-frequencies (or natural frequencies). Substitution of a discrete frequency into Eq. (3.10) allows evaluation of $\{\phi\}$ if 1 DOF is assigned an arbitrary magnitude. The column vector $\{\phi\}$ is thus a normalized eigenvector.

Individual terms in $\{\phi\}$ may be visualized as vibration displacement extremes along the structure as it vibrates sinusoidally at a specific natural frequency. Structure natural frequencies are useful to know so that applied time-varying loads do not result in resonance. Also, the eigenvalues and eigenvectors are useful in applying an arbitrary time history to a structure (see p. 35).

Mass distribution is important in a dynamic vibration analysis. The default in most finite element programs is to lump element mass into adjoining grid points. This is called a lumped mass approach and results in a diagonal mass matrix (see p. 37). An alternative consistent mass approach [2] uses the same element shape functions (p. 14, Chapter 2) used for element stiffness. A consistent mass matrix will have off-diagonal terms as does the stiffness matrix.

Frequency Response

Let the general structural response equation have zero nonlinear loads, zero geometric stiffness, and be subjected only to sinusoidal time load history. However, each applied load may be at an arbitrary phase angle. Equation (3.6) then becomes [5]

$$[-M\omega^2 + i\,B\,\omega + K]\,\{u\} = [P] \qquad (3.11)$$

where i = unit imaginary number for 90° phase relationship.

The most common engineering application of Eq. (3.11) is to apply steady-state sinusoidally varying loads at several points on a structure (perhaps out of phase) and determine structure response over a frequency range of interest. Structure displacement (and stress) response will be complex; i.e., out-of-phase relationships exist. A direct or modal solution of Eq. (3.11) is quite similar in concept to transient analysis, which is considered on page 36.

Nonlinear Statics

Nonlinear structural analysis must be considered if large displacements occur with linear materials (geometric nonlinearity), small displacements occur with nonlinear stress-strain relationships for structural materials (material nonlinearity), a combination of large displacements and nonlinear stress-strain effects occurs (geometric and material-nonlinearity), or gaps appear and/or sliding occurs between mating components during load application or removal.

Geometric Nonlinearity

Structures often undergo large displacements with linear stress-strain relationships. Examples of geometric nonlinear problems are shown in Figure 3.5. The horizontal wire has no lateral stiffness using small displacement theory finite element analysis. However, the wire will carry transverse load due to

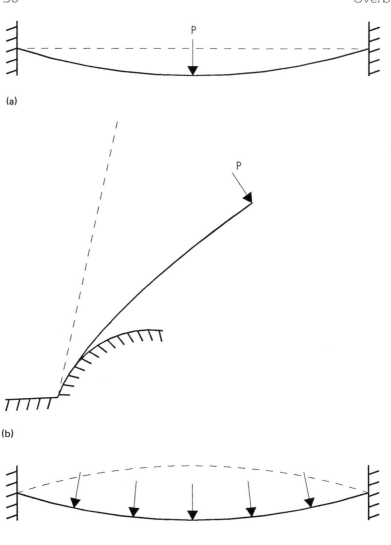

(a)

(b)

(c)

Figure 3.5 Geometric nonlinear examples. (a) Taut wire, lateral load. (b) Large displacement, variable support. (c) Pressurized diaphragm, snap-through.

tension as it is deformed. This effect is evident only where large displacement effects are considered.

A flat plate structure loaded normal to its surface is a candidate for geometric nonlinear analysis if the plate displacement is more than about one-half the plate thickness.

The large rotation of a cantilever beam shown in Figure 3.5b represents both large displacement (and rotation) and closing gaps during loading. The snap-through diaphragm (shallow shell) in Figure 3.5c represents a class of geometric nonlinear problems in which several quasistationary solutions exist along with a sudden instability at the snap-through condition.

Figures 3.5b and 3.5c show the effect of follower forces. A point load normal to the cantilever beam at the free end will change direction as the beam deforms. The pressure acting normal to the snap-through diaphragm surface changes direction between the two positions shown on the figure.

Convective Element Coordinates

A geometric nonlinear analysis has by definition large changes in geometry and thus requires that the equilibrium equations be expressed in the deformed state [2,7]. The updated Lagrangian approach involves a local coordinate system that moves with the element as it undergoes rigid-body motion on the deforming structure. Strains and rotations within the convected element coordinate system (see pp. 43 and 45) are small enough that linear material stress-strain relationships are valid. Thus, total displacement from an applied load on the structure is obtained by a series of displacement increments with the local element coordinate system updated after each.

Newton-Raphson Iteration

The Newton-Raphson method of obtaining roots to polynomial equations numerically has been used successfully for many years. Here it will be demonstrated on a single-DOF structure to obtain equilibrium displacement for an applied load where the load-displacement curve is nonlinear.

If a single-DOF structure is in equilibrium, the applied grid point load balances the element force. If element force is a nonlinear function of displacement, a trial displacement will result in a force imbalance R between applied load P and element force F:

$$R = P - F \tag{3.12}$$

The objective is to reduce R to zero.

In Figure 3.6a, the single-DOF structure is assumed to have zero force

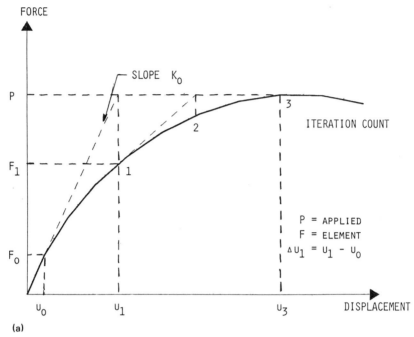

FORCE

SLOPE K_0

P

ITERATION COUNT

F_1

1

2

3

P = APPLIED
F = ELEMENT
$\Delta u_1 = u_1 - u_0$

F_0

u_0 u_1 u_3 DISPLACEMENT

(a)

Figure 3.6 Nonlinear load displacement, iterative solutions. (a) Newton-Raphson method. (b) Modified Newton-Raphson method.

imbalance with element force F_o in balance with an applied load at a displacement u_o. The objective is to find u_3 for applied load P.

A Taylor expansion of $f(u)$ in Figure 3.6a, truncated to the first derivative, at the intial equilibrium point is

$$f(u_o + \Delta u_1) = F_o + \frac{df}{du}\bigg|_o (\Delta u_1) \tag{3.13}$$

From page 19, the derivative term is K_o or the initial stiffness matrix. Hence, the desired incremental displacement for applied load P is

$$K_o (\Delta u_1) = P - F_o \tag{3.14}$$

Figure 3.6a shows a force imbalance R_1 at point 1. Hence, at least one further iteration will be required to achieve equilibrium. A logical iteration procedure [2] would be

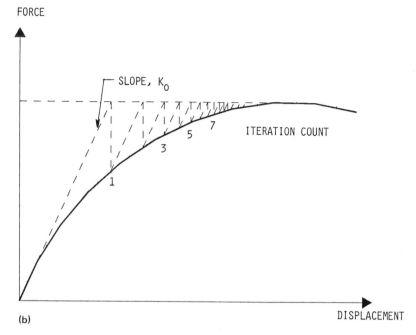

FORCE

SLOPE, K_o

ITERATION COUNT

(b)

DISPLACEMENT

Figure 3.6 Continued.

obtain $u_1 = u_o + \Delta u_1$
compute K_1 and F_1 using u_1
calculate Δu_2 as in Eq. (3.14) using the updated results
continue to iterate until equilibrium is reached where u_3 is the correct
  displacement.

A costly part of Newton-Raphson iteration is updating the stiffness matrix
at each iteration. This leads to the modified Newton-Raphson method shown
in Figure 3.6b. As the figure shows, each iteration has the same slope K_o,
and more iterations are required than for the Figure 3.6a technique.

The force-displacement curve in Figure 3.6 is representative of a softening
structure where K decreases with increasing displacement. A hardening structure with K increasing with displacement may lead to convergence problems,
especially using the modified Newton-Raphson method. Neglecting geometric
stiffness components of the tangent stiffness matrix often will improve convergence [7].

Material Nonlinearity

Material nonlinearity finite element analysis involves nonlinear stress-strain relationships, as shown in Figure 3.7. The figure shows two concepts of material nonlinear analysis. In Figure 3.7a an element is loaded along a nonlinear stress-strain curve and returns to a zero stress-strain state during unloading. There is a known relationship between stress and strain even though an iterative solution procedure would be required (see preceding section). In Figure 3.7b, on the other hand, the element is loaded elastically to a yield stress, further loaded along a smaller-slope stress-strain curve to maximum load, and then unloaded elastically to a state of zero stress but with a residual strain. This type of analysis requires incremental load application and iterative procedures in problem solution.

Material nonlinear analysis involving plasticity (Figure 3.7b) is an extremely complex subject and is discussed in detail elsewhere [2,7]. The most popular solution method is based on incremental or flow theory, wherein increments

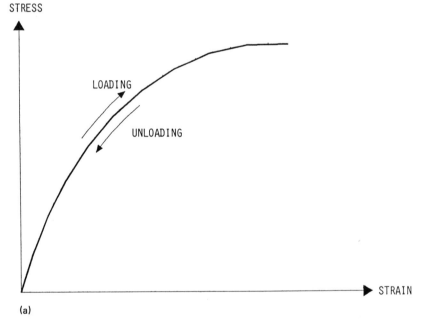

(a)

Figure 3.7 Material nonlinear static analysis, constitutive relationships. (a) Nonlinear elastic. (b) Nonlinear plastic.

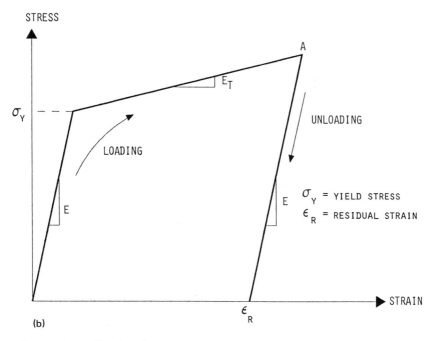

Figure 3.7 Continued.

of stress are related to increments of strain. Two- and three-dimensional elements require yield criteria formulae relating an effective stress at yield to the principal stresses in the finite element. In addition, hardening rules are required to predict change in yield stress if an unloaded structure is reloaded. For example, the Figure 3.7b stress-strain curve would show yield at point *A* if the unloaded structure were reloaded using an isotropic hardening rule.

In general, displacement in a plastic analysis of material nonlinear structures is load-path-dependent, and the iterative Newton-Raphson technique is used in problem solution. The load should be applied incrementally, and subincrements (under program logic control) are usually employed to evaluate particular elements that may exceed yield within a load step increment.

Linear Transient Response

The general structural analysis Eq. (3.6) with linear (Hooke's law) material and without nonlinear loads $\{N\}$ may be written as

$$[Mp^2 + Bp + K]\{u(t)\} = \{P(t)\} \tag{3.15}$$

where $p = d/dt$ and it is emphasized that the structural displacements and applied loads depend on time t.

The preferred method of solving Eq. (3.15) is to select a small time step Δt and, starting at a known initial condition t_0, solve for $\{u\}$ at successive increments of Δt. An implicit method of numerical integration can be made unconditionally stable regardless of the size of Δt [2].

One method to solve Eq. (3.15) uses the Newmark β method where the displacement vector $\{u\}$ at $t + 2\Delta t$ is found in terms $\{u\}$ and $\{P\}$ at time $t + \Delta t$ and t:

$$[A_1]\{u_{n+2}\} = \frac{1}{3}\{P_{n+2} + P_{n+1} + P_n\} + [A_2]\{u_{n+1}\} + [A_3]\{u_n\} \tag{3.16}$$

where

$$[A_1] = \left[\frac{M}{\Delta t^2} + \frac{B}{2\Delta t} + \frac{K}{3}\right]$$

$$[A_2] = \left[\frac{2M}{\Delta t^2} - \frac{K}{3}\right] \tag{3.17}$$

$$[A_3] = \left[\frac{-M}{\Delta t^2} + \frac{B}{2\Delta t} - \frac{K}{3}\right]$$

Direct Transient Response

Transient response analysis consists of assembling the dynamic equations, solving these equations for displacement by Eq. (3.16), and recovering time variation of displacements and stresses. Assembly and data recovery increase linearly with model DOF, but equation solution varies as the square or cube of DOF [5]. Hence, a model with more than a few hundred DOF and several hundred time steps can become quite expensive.

Direct transient response solves Eq. (3.16) using all model DOF. Some transient analysis programs allow the user to input nonzero initial conditions for direct transient analysis. Solution time increases dramatically with model size for models with more than a few hundred DOF. CPU time for a given model solution is nearly directly proportional to the number of time steps [6].

Modal Transient Response

Modal transient response analysis is an alternative to direct transient analysis. The technique consists of the following steps:

Extract model real eigenvalues (p. 28) and eigenvectors over a frequency
 range somewhat greater than the highest frequency in the excitation load.
Integrate Eq. (3.15) using each natural frequency (and eigenvector) as a DOF
 and choosing a suitable small time step to sample the time-varying applied
 load and to display structure response.

The modal method may decouple the equations of motion [2,5,6] such that
the contribution of each structure natural frequency to total structure response
is computed separately. However, even if the equations are coupled (see next
section), the solution time will generally be reduced compared to direct transient analysis because the structure is represented by fewer DOF.

A 3-DOF undamped spring-mass system will illustrate rigid body motion,
real eigenvalue extraction, normalized eigenvectors, and generalized coordinates as used in modal transient analysis.

Figure 3.8a shows three masses joined by two springs and free to translate
horizontally, which indicates a rigid body (zero frequency) vibration mode.
The equations of motion in differential form become

$$\ddot{u}_1 + u_1 - u_2 + 0\,u_3 = 0$$

$$2\,\ddot{u}_2 - u_1 + 2.5\,u_2 - 1.5\,u_3 = 0 \tag{3.18}$$

$$3\,\ddot{u}_3 + 0\,u_1 - 1.5\,u_2 + 1.5\,u_3 = 0$$

These equations take the following matrix form:

$$\begin{bmatrix} 1 & 0 & 0 \\ 0 & 2 & 0 \\ 0 & 0 & 3 \end{bmatrix} \begin{Bmatrix} \ddot{u}_1 \\ \ddot{u}_2 \\ \ddot{u}_3 \end{Bmatrix} + \begin{bmatrix} 1 & -1 & 0 \\ -1 & 2.5 & -1.5 \\ 0 & -1.5 & 1.5 \end{bmatrix} \begin{Bmatrix} u_1 \\ u_2 \\ u_3 \end{Bmatrix} = \begin{Bmatrix} 0 \\ 0 \\ 0 \end{Bmatrix} \tag{3.19}$$

Real eigenvalue analysis assumes all parts of the structure vibrate sinusoidally
(p. 28). Let $u_1 = Ui \cos \omega\, t$, where Ui is the maximum displacement of the
*i*th mass. Equation (3.10) using $\{u\}$ instead of $\{\phi\}$ yields

$$\begin{bmatrix} (1 - \lambda) & -1 & 0 \\ -1 & (2.5 - 2\lambda) & -1.5 \\ 0 & -1.5 & (1.5 - 3\lambda) \end{bmatrix} \begin{Bmatrix} U1 \\ U2 \\ U3 \end{Bmatrix} = 0 \tag{3.20}$$

where $\lambda = \omega^2$.

The matrix in Eq. (3.20) may be expanded about the first row to yield

$$(1 - \lambda) \left[\left(\frac{5}{2} - 2\lambda \right) \left(\frac{3}{2} - 3\lambda \right) - \frac{9}{4} \right] - \left(\frac{3}{2} - 3\,\lambda \right) = 0 \tag{3.21}$$

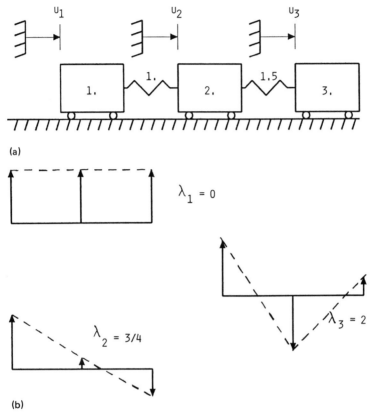

Figure 3.8 Three-DOF structure, vibration frequencies and eigenvectors. (a) Physical arrangement. (b) Eigenvector graphic display.

and factored

$$\lambda \left(\frac{3}{4} - \lambda \right) (2 - \lambda) = 0 \tag{3.22}$$

By inspection, the roots of Eq. (3.22) yield eigen-frequencies $\lambda_1 = 0$, $\lambda_2 = 3/4$, and $\lambda_3 = 2$.

The eigenvectors for the Figure 3.8a model are found by successively substituting the three eigenvalues into the first and third equations of Eq. (3.20):

$$\lambda = 0 \quad U1 = U2 \quad 1.5U2 = 1.5U3 \quad \phi^T = \begin{bmatrix} 1 & 1 & 1 \end{bmatrix}$$

$$\lambda = \frac{3}{4} \quad \frac{U1}{4} = U2 \quad 1.5U2 = -0.75U3 \quad \phi^T = \begin{bmatrix} 1 & \frac{1}{4} & -\frac{1}{2} \end{bmatrix}$$

$$(3.23)$$

$$\lambda = 2 \quad U1 = -U2 \quad 1.5U2 = -4.5U3 \quad \phi^T = \begin{bmatrix} 1 & -1 & \frac{1}{3} \end{bmatrix}$$

These eigenvectors are sketched in Figure 3.8b. The zero-frequency rigid body mode shows each of the three masses translating unit distance in the same direction.

The three eigenvectors may be combined to form a modal matrix:

$$[\phi] = [\phi_1 \ \phi_2 \ \phi_3] = \begin{bmatrix} 1 & 1 & 1 \\ 1 & .25 & -1 \\ 1 & -.5 & .33 \end{bmatrix} \qquad (3.24)$$

where each column represents one of the three normalized eigenvectors. Hence, the modal matrix of eigenvectors relates physical displacement vector $\{u\}$ to a generalized coordinate vector $\{\zeta\}$:

$$\{u\} = \begin{Bmatrix} u_1 \\ u_2 \\ u_3 \end{Bmatrix} = [\phi] \{\zeta\} = \begin{Bmatrix} \zeta_1 + \zeta_2 + \zeta_3 \\ \zeta_1 + .25\zeta_2 - \zeta_3 \\ \zeta_1 - .5\zeta_2 + .33\zeta_3 \end{Bmatrix} \qquad (3.25)$$

Natural frequencies of a large model with N DOF may be truncated to M of the lower frequencies, where $M << N$. Then M generalized coordinates represent the structure. Integration of Eq. (3.15) by using Eq. (3.16) would reduce transient response solution cost over a direct transient response analysis, because the $u(t)$ are replaced by fewer $\zeta(t)$. Data recovery for grid point displacements uses generalized coordinates $\{\zeta\}$ solution results and eigenvectors $[\phi]$ as in Eq. (3.25).

Structural Damping

Structural transient response and/or frequency response requires damping to simulate actual response. Viscous dampers can simulate familiar components such as automotive or aircraft shock absorbers. Also, most metals exhibit some structural damping, due to relative motion within the material on an atomic level. A third type of damping, modal damping, may be imposed in modal transient and/or frequency response solutions. Viscous damping is proportional to the relative velocity across the viscous damper and implemented by specifying a damping coefficient. Modal and structural damping is specified as a fraction of critical damping. Only modal damping decouples the equations of motion expressed in generalized coordinates. Any small amount of viscous or structural damping causes coupling of the equations [5].

Nonlinear Transient Response

Nonlinear transient response [2,7] involves a combination of material nonlinear static analysis (p. 34) and linear transient response solution technique, such as the Newmark β method Eq. (3.16). There is no guarantee of unconditional convergence in that the Newton-Raphson interation at a particular time step may fail to converge and the time integration of the dynamic equations may diverge. Loads are usually applied in increments, and reducing the time step enhances convergence. Some finite element computer programs automatically subdivide the time step in an attempt to achieve convergence.

STRUCTURAL ANALYSIS PROGRAMS

The engineer has access to many finite element computer programs. These programs may be used on a computer time-sharing service, or computer-program vendors will install the software on the engineers' minicomputer or microcomputer. Some limited-capacity finite element programs operate on a personal computer.

Three world-class finite element structural analysis programs are ABA-QUS, ANSYS, and MSC/NASTRAN. All three have a wide range of similar analysis capabilities. Each has an appropriate user's manual [8–10] and other documentation including specific analysis techniques in handbooks [5,11,12], theoretical manuals, and verification or demonstration problem manuals. MSC/NASTRAN, in particular, has an application manual in which new analysis techniques and elements are explained in a timely manner to the program users. All three programs have frequent updates, telephone hotlines to answer user questions, and a full-time professional staff.

Differences in solution approach, element definition, and solution economy are topics more fully considered as appropriate in the discussion below. Unless stated otherwise, this chapter will consider only the MSC/NASTRAN finite element computer program because of the author's familiarity with and access to this program.

COORDINATE SYSTEMS

As mentioned in Chapter 2 (p. 8), finite element models are defined by grid points located in three-dimensional space. Location of these grid points is defined by coordinate systems. Also, grid point displacements and other properties may for convenience require a different coordinate system from that used to locate the grid points. A summary [7] of coordinate systems used in finite element analysis would include

Fundamental
Local
Element
Displaced
Material

Fundamental Coordinate System

The three finite element programs discussed above are based on a right-hand Cartesian coordinate system. Local coordinate systems must refer to the fundamental coordinate system or another local coordinate system based on the fundamental system.

The default of the three finite element programs is to locate, constrain, and output displacements in the fundamental coordinate system. Also, many finite element model generation programs [13] default to the fundamental coordinate system. The coordinate system identification number zero is reserved for the fundamental system.

Local Coordinate System

Definition of local rectangular and cylindrical coordinate systems is shown in Figure 3.9. Spherical coordinate systems may also be defined. The location technique involves assigning an integer for identification and defining an origin (point A), a z axis (along a vector from point A to point B), and the rectangular XZ plane or cylindrical azimuthal plane (the plane containing the vector from point A to point C). These three definition points may be expressed in reference system coordinates or as grid points defined in another coordinate system.

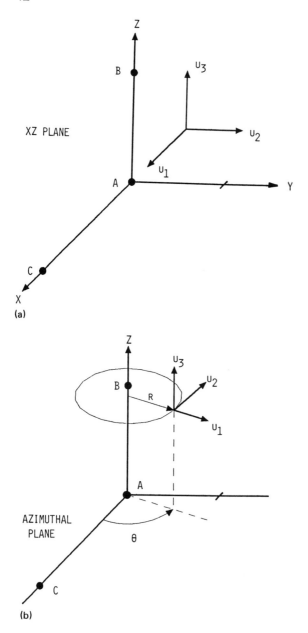

Figure 3.9 Local coordinate systems. (a) Rectangular system. (b) Cylindrical system.

Defining local coordinate systems in terms of other local coordinate systems (eventually referencing the fundamental system) is called nesting.

Local coordinate systems used to locate grid points are called local by all three programs. Local coordinate systems used to output displacement results (and specify constraints) are called transformed, nodal, and displacement in ABAQUS, ANSYS, and MSC/NASTRAN, respectively.

Element Coordinate Systems

Each element has an element coordinate system (see p. 46) used to output stress, shears, and moments, and to input section properties for one-dimensional elements. These coordinate systems are defined in terms of element grid point connectivity for two- and three-dimensional elements and in terms of connectivity and property-orientation vectors for one-dimensional beam elements.

Displaced Coordinate System

Large-displacement analysis (p. 29) may use a convective or displaced coordinate system to account for element rigid body translation and rotation on a deforming structure. This analysis technique allows for large geometric displacement and yet small displacement elastic deformation of the element in a displaced orientation.

Material Coordinate Systems

Two-dimensional and three-dimensional elements allow the user to input a material coordinate system to orient orthotropic or anisotropic material axes (see p. 51) within the element. This is accomplished on two-dimensional element connectivity cards. Local coordinate systems may be referenced on three-dimensional element property cards (see p. 50).

ELEMENT TYPES

The three structural analysis finite element programs cited above differ considerably in element definitions for the structural analysis problem types discussed on page 25. The ABAQUS and ANSYS programs have elements classified, for example, as two-dimensional elements strictly for planar analysis, whereas other two-dimensional elements may also be used in three-dimensional models. These two programs have many elements suitable only for structural analysis, thermal analysis, or material nonlinear analysis. ANSYS has a significant number of combination elements. Table 3.1 summarizes available classes of ABAQUS and ANSYS elements.

Table 3.2 summarizes MSC/NASTRAN elements, which will be discussed in detail. These elements are generally used in either one-, two-, or three-dimensional models. The user must constrain unneeded grid point DOF when solving planar models. Also, most elements accommodate structural and thermal analysis, and many are suitable (with some restriction on edge grid points) for geometric and material nonlinear analysis.

Finite elements are described by a connection card list of grid points and reference to a property card. The connectivity list usually orients the element coordinate system and may orient the material coordinate system. MSC/NASTRAN prefixes a C to the element names in Table 3.2 for the connection card. Also, a vector to orient one-dimensional element cross-sectional axes

Table 3.1 ABAQUS and ANSYS Finite Elements

Elements	ABAQUS[a]	ANSYS[a]
Total elements	98	71
Structural[b]	54	45
Heat transfer	19	18
Combination[c]	24	6
Miscellaneous	1	2

[a]Data for ABAQUS [8] and ANSYS [9].
[b]Includes nonthermal gap (interface elements).
[c]Combination solutions: displacement/thermal, displacement/pore pressure, thermal/electric resistance.

Table 3.2 Selected MSC/NASTRAN Structural Elements

Element dimension	Elements[a]
0	CONM, ELAS, GENEL, MASS, RBE
1	BAR, BEAM, BEND, ROD
2	QUAD, SHEAR, TRIA
3	HEXA, PENTA, TETRA, TRIAX

[a]Detailed descriptions of these elements may be found in Ref. [3].

is included on the connection card. Many connection cards may reference a single appropriate property card.

A finite element property card provides additional geometric information, nonstructural mass distribution, and references a material card. Geometric data include element thickness for two-dimensional elements and cross-section properties for one-dimensional elements. Property cards are formed by pre-fixing a *P* on the names of Table 3.2 one-dimensional elements and the SHEAR element. Other two-dimensional elements reference a PSHELL property card and most three-dimensional elements reference a PSOLID property card.

The most commonly used isotropic material card would specify elastic modulus, shear modulus, Poisson's ratio, density, thermal expansion coefficient, reference thermal expansion temperature, and structural damping for dynamic analysis. Various forms of *MATi* (*i* being an integer) acronyms describe isotropic, orthotropic, and anisotropic material cards (see p. 51). Also, the material may be made temperature dependent by referencing *MATTi* cards, which reference appropriate tables containing temperature-property pairs. It is recommended that isotropic materials not use a shear modulus entry, which is internally calculated from elastic modulus and Poisson's ratio.

One-Dimensional Elements

Truss Element

The most simple one-dimensional truss element (see p. 19) has pinned ends and supports only extensional loads. It is a straight prismatic (constant cross-section) element, and the element *x* axis passes through the element centroid from the first grid point to the second grid point on the element connection card.

The ROD element also has torsional stiffness. Its property card specifies cross-section area, torsion stiffness constant, nonstructural mass per unit length, and torsional stress recovery distance from the element axis. Extensional or torsional stiffness may be deleted by omitting the appropriate entry on the property card. Axial stress or force and torsional stress are output by the element.

Truss elements find application in modeling:

Pin-ended truss or space-frame structures
Reinforcement for two-dimensional elements not requiring bending stiffness
Torque tubes or torsional springs
Cables that are known to only be in tension

Other forms of rod elements represent tubular truss elements or omit a property card for extensional application by including cross-section area on the connectivity card.

Bending Elements

A suitable beam element in a space frame requires at minimum extensional and torsional stiffness as well as bending stiffness and transverse shear flexibility in two orthogonal planes. This describes the MSC/NASTRAN BAR element shown in Figure 3.10a. This prismatic element is connected between two grid points with a coincident centroid and shear center. Hence, it is suitable for most closed-section beams but only an approximation for angles and other open-section structural members. Plane cross-sections remain plane during deformation.

Two orthogonal bending planes shown in Figure 3.10a are specified by a vector from end A and the element X-axis vector. This is called plane 1 and contains the element Y-axis orthogonal to x. Plane 2 is orthogonal to plane 1 and contains the element Z-axis. The centroidal axis end points A and B may

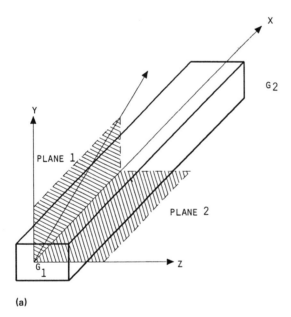

(a)

Figure 3.10 Element coordinate systems. (a) One-dimensional bar. (b) Two-dimensional quadrilateral plate. (c) Three-dimensional pentrahedron.

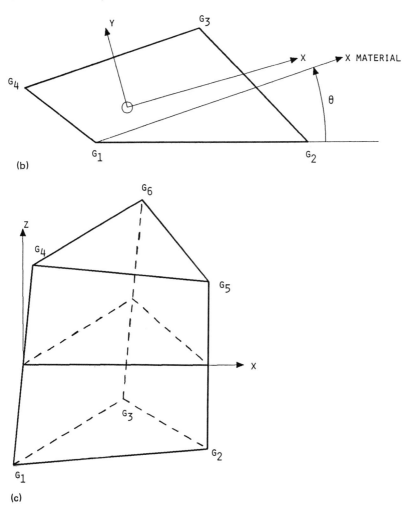

(b)

(c)

Figure 3.10 Continued

be offset from the two grid points defining the BAR element. Planes 1 and 2 must be oriented along principal axes of the element cross-section to avoid specifying a product of inertia, which may be input on the property card.

The BAR element property card specifies cross-section area, torsional stiffness, and bending moments of inertia in planes 1 and 2. Four stress recovery points are defined at each end normal to the two bending planes, and nonstructural mass distribution is specified. The property card allows removal of stiffness connection at each end between the neutral axis and defining grid points. These joints are defined in the element coordinate system and allow for pinned or ball-joint applications. The element outputs axial force and stress along with bending moments, transverse shear, and bending stress in the two planes at both ends.

The BEAM element separates the neutral axis, shear center axis, and nonstructural mass axis. Bending planes are defined as for the BAR. The element is straight but need not be prismatic. Section properties may be arbitrarily defined at each end and up to nine intermediate points. The cross-section may warp, and the effect on torsional stiffness is accounted for by using a scalar point—a DOF not physically located in space.

The BEND element simulates a curved beam of arbitrary cross-section or a curved pipe. A circular arc is formed between two grid points to yield extensional and torsional stiffness along with bending stiffness and transverse shear flexibility in two orthogonal planes. The curved-pipe option accounts for internal pressure stiffening as well as curvature effect on bending stiffness and stress. Forces and moments are output at each end as well as axial stress at four cross-sectional points.

BAR, BEAM, and BEND elements contribute 6 DOF stiffness to grid points unless pinned end options are used.

Two-Dimensional Elements

Two-dimensional elements tabulated in Table 3.2 are typical of modern elements in structural analysis programs. Most are isoparametric [2,3], which may be briefly described as an element with one describing polynomial (see p. 12, Chapter 2) that relates element internal coordinates to grid point coordinates and element internal displacement to grid point displacement.

Bending and Membrane Elements

The linear strain quadrilateral element shown in Figure 3.10b and tabulated in Table 3.2 is considered to be the best two-dimensional element in the MSC/NASTRAN library [11]. The element has either four connecting grid points (QUAD4) or an optional four edge grid points (QUAD8). The element has no elastic resistance to the rotational DOF normal to the surface. Hence, the element is attached to only 5 of the 6 DOF at the grid point. Another quadrilateral element at slight out-of-plane angle to the Figure 3.10b element will contribute stiffness to the rotational DOF normal to this single element.

Element coordinate systems for these elements are defined in terms of the connecting grid points. The QUAD4 in Figure 3.10b, for example, has an element coordinate system origin where the two diagonals intersect with the element positive X-axis in the general direction from G_1 to G_2 bisecting the angle between the diagonals. The element Y-axis is orthogonal to the X-axis in the general direction from G_1 to G_4. The Z-axis is normal to the element as defined by the right-hand rule. The element coordinate system is used to output stress, force, and moments. An element material coordinate system is oriented from the $G_1 - G_2$ edge.

If edge grid points are used in QUAD8, they need not be on straight lines to corner grid points. However, a midedge grid point is desirable, and a grid point at an edge quarter-point results in numerical difficulty [2]. Midedge grid point QUAD8 elements are useful in mesh refinement areas of a model.

The triangular elements consist of TRIA3 or TRIA6 with edge grid points. These are constant strain elements and are recommended [11] only if quadralateral elements are inconvenient to model structure geometry. The element X-axis is parallel to the $G_1 - G_2$ edge and the Y-axis orthogonal in the $G_1 - G_3$ direction. These elements also have no elastic resistance in rotation normal to the surface, and edge grid points should be located at midedge.

Element properties (PSHELL) specify element thickness and material card identification (MID). Up to four material selections are possible (either different or identical). Each contributes stiffness or coupling of element deformation:

MID1—membrane stiffness
MID2—bending stiffness
MID3—transverse shear stiffness
MID4—membrane and bending coupling

A homogeneous plate element references a property card with MID_1—MID_3. The use of MID_4 is recommended only by very experienced analysts. Honeycomb material properties may also be specified for triangular and quadrilateral elements.

Shear Panel Element

Analysis of thin-skinned curved structures (such as aircraft) has traditionally used shear panel and truss elements. The quadrilateral SHEAR element (PSHEAR property card) supports extensional force at grid points and shear stress within the element. Reinforcing ROD elements are used to carry extensional load.

Axisymmetric Elements

The TRIAX element listed in Table 3.2 is a triangular cross-section axisymmetric element with three corner and three midedge grid points and is called TRIAX6. The element is described in the XZ basic coordinate system plane with the Z-axis as the axis of revolution. Radial and axial loads applied are assumed axisymmetric, and the axisymmetric elements may not be used with other elements. The ANSYS program, on the other hand, supports an axisymmetric element that may be loaded (harmonically) in the rotational direction and the element may be combined with other elements.

Solid Elements

Figure 3.10c shows the isoparametric PENTA element with six corner grid points. The element X-axis is directed by $G_1 - G_2$ connectivity with the Z-axis normal to the midedge plane and the Y-axis in the midedge plane defined by the right-hand rule. The element may have up to nine edge grid points located as near midedge as possible. This element is recommended [11] for thick and reasonably thin shell modeling with the triangular faces on the exposed surfaces.

The six-faced, eight-corner-grid-point HEXA element may have up to 12 edge grid points, preferably at midedge. The definition of element coordinates is more complicated than for the PENTA element, but the X-axis is nearly parallel to the $G_1 - G_2$ connectivity.

The four-faced TETRA element has four corner grid points and up to six on edges, preferably located near midedge. The element coordinate system is located at G_1 with a $G_1 - G_2$ X-axis. The TETRA is recommended [11] to be used only to fill in gaps left in a solid element model using HEXA and PENTA elements. Otherwise fine mesh sizes are recommended in areas of high stress gradients.

The HEXA, PENTA, and TETRA elements reference a PSOLID property card. This card selects the material coordinate system, which may be the element coordinate system or a defined system for selection of stress output format. Solid elements only contribute translational stiffness to grid points.

Zero-Dimensional Elements

Table 3.2 lists several elements classed as zero dimensional. The mass element (CONM2) concentrates mass and/or rotational inertia at or offset from grid points. Hence, a CONM2 will show translational inertia in three orthogonal directions when subjected to body-force loads (see p. 53). A CMASS element, on the other hand, specifies a scalar mass or inertia for a single DOF.

The ELAS element defines a spring rate between two DOF. Care must be taken in all finite element programs to have the two connected translational DOF coaxial or spurious support reactions may result.

The GENEL element allows the program user to specify a stiffness matrix of an undefined substructure to a number of grid points on a defined structure. This stiffness matrix may have been obtained from other analysis or experimental data.

Rigid body elements (RBE in Table 3.2) are useful in forcing several grid points to displace in unison as part of a rigid body. These are not elements, which have mass and stiffness, but they are formed from program-generated multipoint constraints (see p. 54). Several variations of rigid body elements meet particular modeling requirements.

MATERIAL PROPERTIES

The three major structural analysis programs described above all have the following material properties available in at least some elements:

Linear isotropic
Linear anisotropic
Linear orthotropic
Nonlinear isotropic
Temperature-dependent isotropic

These terms may be concisely defined [14] as follows:

Isotropic material properties, including strength and stiffness, are independent of loading direction.

Anisotropic material properties are dependent on loading direction, producing different stress and strain with a change in loading.

Orthotropic material properties are anisotropic in which planes of extreme values are orthogonal.

Polar orthotropic material properties, such as wood, are orthogonal in cylindrical coordinates.

Linear Isotropic Material

The most useful material properties for a wide range of structural analyses involve linear isotropic materials. The defined material constants are elastic modulus, shear modulus, Poisson's ratio, and thermal expansion coefficient. All properties are constant during the entire loading and/or unloading of the structure. The material axis may be rotated with respect to the element coordinate system for two-dimensional elements (see Figure 3.10b). The shear modulus G may be internally computed from elastic modulus E and Poisson's ratio.

Linear Anisotropic Material

Anisotropic material analysis may require that a 6×6 symmetrical material matrix and a six-row column vector of thermal expansion coefficients be defined. Individual terms in the 6×6 material matrix are defined [3] in terms of three orthogonal values for elastic modulus, shear modulus, and Poisson's ratio.

Two-dimensional elements require a 3×3 symmetrical material matrix and a three-row column vector of thermal expansion coefficients. Formulae for calculation of three-dimensional and two-dimensional anisotropic material matrices in terms of elastic moduli, shear moduli, and Poisson's ratios for elements are available [3].

Linear Orthotropic Material

An orthotropic material is a special case of the more general anisotropic material capability. The two-dimensional orthotropic material matrix involves four independent constants [11].

Nonlinear and Temperature-Dependent Isotropic Material

The stress-strain curves shown in Figure 3.7 are an example of nonlinear isotropic material elastic modulus. These data are generally input in tabular format of stress-strain pairs. The ANSYS program allows input of a family of curves that are temperature dependent.

LOADS

Static Loads

Structural analysis involves application of loads at grid points. However, as a user convenience distributed load and initial strain may be applied to one-dimensional elements, pressure loads on two- and three-dimensional element surfaces, and gravity, centrifugal load, and thermal strain applied to all elements. These loads are then resolved to connecting grid points for program solution.

Direct Loads and Enforced Displacements

FORCE and MOMENT cards are used to apply forces or moments directly to grid points. The direction of the applied force or moment, for example, is defined by three orthogonal components of a predefined coordinate system or along a vector between two grid points.

Enforced displacements may be defined at grid points using constraint cards (see p. 54) or an enforced displacement (SPCD) that resembles an applied force or moment.

Distributed Loads

The PLOAD*i* (where $i = 1,2,3,4,X$) defines loads along one-dimensional elements or on faces of two- and three-dimensional elements. Considerable generality is available in line load specification, such as concentrated and linearly varying loads along BAR, BEAM, and BEND elements. Element surface pressure load may vary linearly across the face and be specified normal to the face or at an arbitrary direction in model coordinates.

Gravity and Centrifugal Loads

Body-force loads due to gravity (or some other uniform acceleration) may be expressed by using a GRAV card, which defines the gravity coordinate system, magnitude, and directional components. The mass matrix is multiplied by gravity vector components.

The RFORCE card defines an axis of rotation and angular velocity. A centrifugal force resultant equals the product of grid point mass, distance from the spin axis, and angular velocity squared.

Thermal Expansion Load

The most general specification of thermal expansion loads is accomplished by specifying grid point temperatures and a thermal expansion coefficient and reference temperature on the element material card. Other cards specify average temperature and/or temperature gradients along or through cross-sections and thicknesses of one- and two-dimensional elements, respectively.

Dynamic Loads

Frequency response (p. 29) and transient response (p. 35) require frequency variable and time variable loads, respectively. The frequency response load is determined by choosing an amplitude, frequency, or phase angle since all loads are sinusoidal. Transient response loads are time functions. Both types of loads may be classed as enforced loads or enforced motion.

Enforced Load

Enforced loads allow time variation of direct loading at grid points or distributed element loads as used in static analysis (p. 52). Load application procedures accept time-load tabular input or harmonic components [11]. The modal transient method (see p. 36) does not require reanalysis of structure eigenvalues if a new enforced load analysis is desired.

Enforced Acceleration, Velocity, and Displacement

The enforced-motion technique [5] allows input of a time-varying acceleration, velocity, or displacement at grid points. The technique requires that a large mass (say 10^4 times structure mass) be positioned at the grid point prior to real eigenvalue analysis. The applied enforced-motion amplitude is multiplied by a factor equal to the mass. A user-specified code is used to input acceleration, velocity, or displacement. A change in grid point location of enforced motion requires reanalysis of the eigenvalues in a modal transient analysis.

CONSTRAINTS

The stiffness matrix for the three-member truss in Figure 3.3a is singular since it has an order of six and a rank of three [2]. This occurs because the truss is not supported to prevent rigid body motion. The statically determinant truss support in Figure 3.2b requires that grid point displacements $u_1 = u_2 = v_2 = 0$. These three zero displacements are imposed on a finite element model using single point constraints (SPC). The constrained stiffness matrix of the Figure 3.3a truss model would have an order and rank of three. It would be a nonsingular matrix and allow calculation of displacement components v_1, u_3, and v_3 for a load, say, at grid point 3. The constraint forces can then be recovered at the constrained DOF.

   This section discusses single-point constraints, as well as multipoint constraints (MPC), and constraint applications to form rigid elements and partial models of symmetrical structures.

Single-Point Constraints

A single-point constraint (SPC) is an enforced single component of motion applied to grid and/or scalar points. Three translational and three rotational displacements may be specified as real numbers (including zero) at a grid point, and a displacement may be specified at a scalar point. Applications of SPCs in structural analysis include:

Constrained or enforced translations and/or rotations for structure support, and zero motion symmetric and/or antisymmetric boundary conditions in planes of structural symmetry (see p. 56).

Removal of all DOF from unattached grid points and/or scalar points in a model.

Removal of weakly coupled rotational DOF from plate elements and all rotational DOF from grid points attached only to solid elements (see pp. 48 and 50).

SPCs are applied in the grid point displacement coordinate system (p. 43). Hence, if a grid point is to be constrained at an angle with respect to the fundamental coordinate system, a local coordinate system would be employed. Suppose, for example, that a series of truss elements are oriented in a planar analysis as shown in Figure 3.1b to represent a cable in tension. Since the rod has only extensional stiffness, a local coordinate would be employed with the X-axis along the rods. The Y-axis displacement and Z-axis rotation would be constrained at all grid points along the truss-element cable.

An SPC card is used to arbitrarily enforce a DOF at grid or scalar points. The popular SPC1 card enforces selected DOF to zero for many grid or scalar points. Also, an automatic constraint feature (AUTOSPC) may be selected to constrain a stiffness DOF at a grid point having a ratio to the largest stiffness at the grid point less than a specified value, say, 10^{8}. It is recommended that the analyst examine DOF being automatically constrained to ensure that faulty modeling is not present.

Multipoint Constraints

Multipoint constraints (MPC) are used to impose linear relationships between grid and/or scalar point DOF using the following relationship:

$$\sum_i A_i u_i = 0 \tag{3.26}$$

where A is a constant and u is a translational or rotational DOF.

This relationship allows considerable freedom in describing relative motion between DOF. For example, a preload can be imposed on a bolt by defining an MPC relationship such that a scalar point DOF represents the axial-displacement difference between a grid point on the end of a BAR element representing the bolt and another grid point on the bolt-hole structure. The scalar point is constrained (SPC) to a required initial bolt deformation. This preload will be imposed even if other structural deformations tend to increase or decrease the total bolt load, which may be output as axial force in the BAR.

Rigid Elements

Rigid elements (p. 51) are formed using automatic generation of MPC relationships. Rigid elements are processed in the same fashion as user-specified MPC relationships using Eq. (3.26).

Dihedral and Quadrilateral Symmetry

Single-point constraint relationships are useful in solving finite element models of geometrically symmetric structures using partial models. The two most common occurrences are dihedral and rotational symmetry.

Dihedral symmetry (or mirror-image symmetry) is a special case of analysis simplification suitable for structures with geometric symmetry about a centerline. For example, a ladder has geometric symmetry about a centerline running from the base to the top. A railroad car generally has symmetry about the longitudinal and transverse centerline; i.e., this structure possesses "quadrilateral" symmetry.

A dihedral- or quad-symmetry image structure may be modeled by a one-half and one-fourth model, respectively. Effects of stiffness and mass for the other sections are obtained by applying symmetric and antisymmetric boundary conditions to the boundaries between the modeled structure and non-modeled structure. Considerable computer efficiency is obtained with the smaller models, even though two or more finite element solutions must be calculated.

Dihedral image modeling is used by nearly all finite element analysts; the subject is covered in detail in finite element literature [2,11,15]. A brief explanation of symmetric and antisymmetric boundary constraints may be seen for a prismatic beam model shown in Figures 3.11 and 3.12.

Figure 3.11 shows a beam that has dihedral symmetry in the YZ plane at midspan. Symmetric bending and axial forces and a moment are shown along with the resulting beam displacement specially at the plane of symmetry.

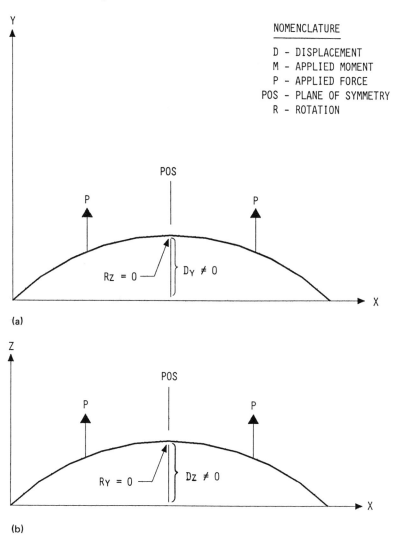

Figure 3.11   Symmetric load motion. (a) Beaming in *XY* plane. (b) Beaming in *XZ* plane. (c) Axial force and moment, *X*-axis.

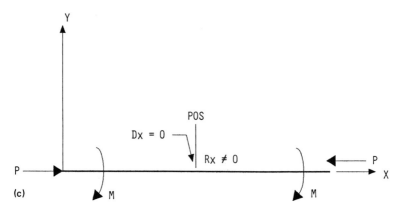

Figure 3.11 Continued

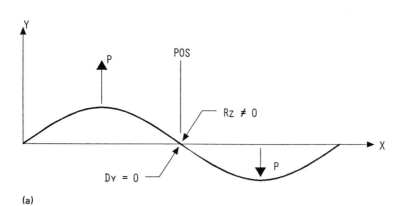

(a)

Figure 3.12 Antisymmetric load motion. For nomenclature, see Figure 3.11. (a) Beaming in XY plane. (b) Beaming in XZ plane. (c) Axial force and moment, X axis.

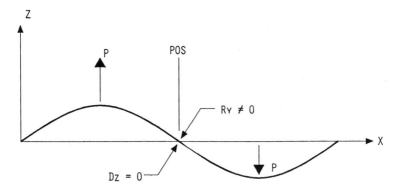

(b)

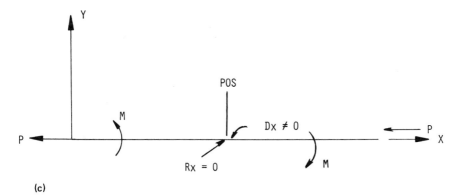

(c)

Figure 3.12 Continued

Figure 3.12 shows the same dihedrally symmetric beam with antisymmetric bending, and axial forces, and moments. Beam displacement differs from the symmetric load case and is identified at the plane of symmetry.

These two loadings of a dihedrally symmetric structure lead to the following rules:

Symmetric boundary conditions (at plane of symmetry) result in zero out of plane motion.

Antisymmetric boundary conditions (at plane of symmetry) result in zero in plane motion.

Figure 3.13 shows how arbitrary unsymmetric loads may be applied to a dihedrally symmetric structure by preparing a half-model, loading the modeled half, solving with symmetric and antisymmetric boundary conditions, and combining the two solutions. Note that a half-load is applied in the modeled half to obtain the correct total loading when the two solutions are combined.

Analysis of a railroad car could use a quarter-model of the structure containing dihedral symmetry about two orthogonal axes. Four finite element solutions with a permutation of symmetric and antisymmetric boundary conditions allow application of an arbitrary loading to any quadrant and data recovery for the modeled and three nonmodeled quadrants.

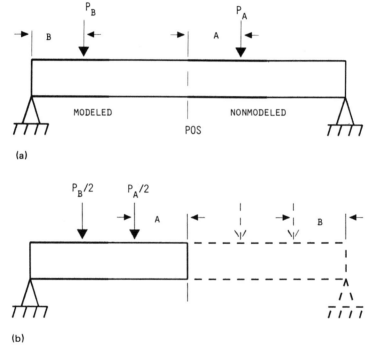

Figure 3.13 Dihedrally symmetric structure, unsymmetric load. (a) Physical arrangement and loads. (b) Half-model, symmetric boundary. (c) Half-model, antisymmetric boundary. (d) Combination of symmetric and antisymmetric solutions.

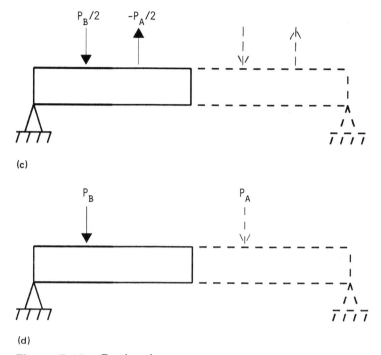

(c)

(d)

Figure 3.13 Continued

It should be noted in the above examples that symmetric supports are assumed in this analysis. Techniques are available to overcome this difficulty, but considerable finite element expertise is required.

Rotational Cyclic Symmetry

Often a structure such as an electric motor rotor or stator consists of cyclic (rotational) symmetry in that the structure is comprised of, say, six rotationally displaced 60° segments. Hence, only one 60° segment is modeled, with stiffness and mass effects of the other five segments obtained by proper application of boundary conditions.

The finite element analyst has some control on rotational cyclic symmetry through multipoint constraints (p. 125, Chapter 5). However, completely arbitrary loading in an arbitrary segment of a rotationally symmetric structure requires special finite element analysis code [11]. The MSC/NASTRAN program has a cyclic symmetry option for static and dynamic solutions. A short-coming in cyclic symmetry is that all segments must have identical support conditions, as was seen above in the dihedral symmetry analysis.

Substructuring

Substructuring is a finite element concept [2,12] in which the structure is subdivided into substructures, which are analyzed independently and combined for a total solution. Stiffness and mass effects are condensed to boundary grid points and result in a "residual" structure. Solution of the residual structure allows for "upstream" data recovery in each substructure.

Recent advanced programs in substructuring [12] have a subset capability of dihedral symmetry without the effort required to specify load case combinations described above. Superelements, a name given to substructures, allow the user to load the images directly by use of residual structure grid points, or to load images by loading the modeled (primary) structure and to position the load by load case control. Also, the modeled structure and mirror images may easily be assigned arbitrary supports by including these grid points in, say, the residual structure which is resolved to change supports. Repeated images such as construction crane boom inserts may be modeled with superelements, whereas cyclic symmetry is unsuitable. Superelement capability also overcomes cyclic symmetry constraint limitations by using rotated images. Cyclic symmetry is also available within the superelement code [12].

RESULTS AND VERIFICATION

The engineer learns (or perhaps memorizes) formulae representing closed-form solutions to engineering problems. Authors [14] warn that every formula has many assumed simplifications and must be used with caution. For example, the beam flexure stress formula assumes that plane sections remain plane, linear stress-strain relationships apply, and strain varies linearly with distance from the neutral plane.

Finite element analysis requires similar precaution from the engineer. This section describes some verification checks that can be made for the finite element model, computer solution, and results interpretation to ensure a satisfactory analysis.

Problem Definition

The first task is to determine what analysis results are required. This involves a decision as to whether static, buckling, natural frequency, frequency response, and/or transient response analyses are needed to convince the designer that the structure will give satisfactory service over its projected life. The advent of reliable nonlinear analysis programs also requires a decision between linear elastic or nonlinear elastic or plastic solution.

A conference between the designer and structural analyst should result in a brief description of the type of analysis, finite element solution technique expected, time duration of the project, and benefits to be derived from the analysis. A one-page summary should be adequate to express these technical goals. An assembly drawing is generally adequate to complete this project brief.

Model Verification

A modern computer-based finite element model generation program will be applied (see Chapter 7) to all but the most simple models. Data may be input manually from detailed drawings or line data from the database used by the designer in his computer-aided design (CAD) function. Many plots from the model generation system and/or preliminary finite element program runs will be required to ensure the following model verification:

Appropriate finite element mesh refinement in areas of expected high-stress gradients.

Correct property cards assigned to elements to correctly model plate thickness, beam section properties, section change fillets, gussets, and other critical structural details.

Rigid elements to represent massive low-stressed unmodeled structure, distribution of pin-bearing loads in holes, uniform force distribution on plate edges, and interconnection between substructures.

Correct material properties for the several components in the structure, including temperature-dependent properties and nonlinear stress relationships if appropriate.

Appropriate structure support or boundary conditions on planes of dihedral symmetry.

Structure estimate and finite element model weight comparison.

Concentrated mass and nonstructural mass distributions, especially in gravity (or centrifugal) static loads and generally for vibration frequency, frequency response, and transient.

Plotting

Clear plots (see Chapter 7) of the undeformed structure and deformed structure due to static load or vibration frequency analysis are essential in model check-out and results verification. Modern video cassette recorders are ideal to animate vibration frequency modes, frequency response effects, and time-varying structure deformation in transient response. Color plots outlining the superimposed deformed and undeformed structure are preferred in reports over the conventional solid line–dashed line plots that are used if color is unavailable.

Solution Verification

Finite element programs display "conditioning and goodness numbers" [3] to indicate possible numerical errors in the solution. For example, MSC/NASTRAN outputs a conditioning number that compares the ratio of diagonal terms in the stiffness matrix to the same diagonal term of an upper triangular factor (see p. 25) resulting during static analysis solution. A ratio greater than 10^8 could indicate possible model support problems or a high stiffness ratio in components at a grid point (see p. 55). The analyst is advised to check areas of the model having a large conditioning number.

A "goodness of solution" number is based on virtual work done by the calculated displacements [3]. This work should be zero at force equilibrium (see p. 11, Chapter 2). This number is called epsilon in MSC/NASTRAN and suggests a numerically correct solution if less than 10^{-5}. Epsilon is calculated for each load case. A large value of epsilon for a load in one particular direction could indicate an error in structure support and/or rigid body interconnection problems at grid points resisting load in that direction.

Results Verification

After making the above modeling and computer solution checks, the results are verified. The most frequent errors result from incorrect structure support or interconnection of components. A summation of translational constraint forces at structure supports is often convenient to check against load components in the same direction. Finally, unusually low stress in a known high-stress area may indicate a need for local finite element mesh refinement.

Structural deformed plots are invaluable in checking whether the solution is reasonable. Frequency response and transient response results may show exceptional displacement due to incorrect viscous or structural damping (see p. 40). Nonlinear instability may require smaller load or time increments.

Experience is invaluable in assessing finite element analysis results, but a careful check on modeling details mentioned above should increase confidence in solution results.

STRUCTURAL ANALYSIS EXAMPLES

As shown on page 25, a large number of structural analysis problem types may be solved using the finite element method. This section will illustrate the following structural analysis problem types:

Linear statics
Linear elastic buckling
Linear real eigenvalues
Linear transient
Geometric-nonlinear statics
Nonlinear-material statics
Nonlinear material transient

Metric units and scientific (exponential) notation will be used to express large or small numbers.

Linear Statics

Figure 3.14 shows a planar frame [1] with two steel built-in beams, welded together at 90°, one of which is free to translate along a surface 30° up from the horizontal. A vertical built-in beam is pinned to the upper frame at the right-angle joint, and a horizontal 1-N force is directed to the right.

Figure 3.15 shows this linear static analysis utilizes three BAR elements (p. 46), since it is assumed that closed-section members are used with coincident centroid and shear center axes. Element 2 is pinned in the element coordinate system at the end joining the welded members. Each BAR cross-section area is 0.001 m^2 and second moment of inertia 0.025 m ** 4. Steel elastic modulus is 2.07E11 N/m^2.

All grid points are located in the fundamental coordinate system, but a local coordinate system with an X'-axis oriented 30° to the fundamental X-axis is specified as a displacement coordinate system at grid point 4. The Y'-component is constrained at grid point 4, since single-point constraints are specified in the displacement coordinate system. Grid points 1 and 3 are fully constrained in all 6 DOF to represent the built-in support. Grid points 2 and 4 are constrained in Z translation and X and Y rotation for a planar analysis. Hence, the frame becomes a 5-DOF model.

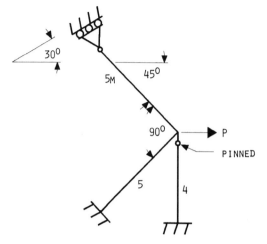

Figure 3.14   Example frame, physical arrangement.

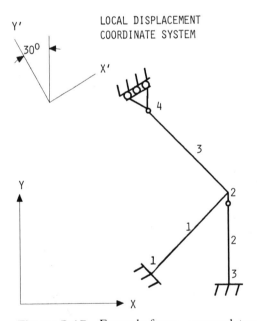

Figure 3.15   Example frame, nomenclature.

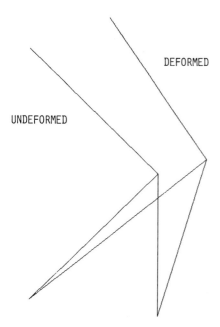

DEFORMED

UNDEFORMED

Figure 3.16 Example frame, linear static deformation.

Figure 3.16 shows the deformed and undeformed frame. As expected, the frame deforms in the direction of the load, and grid point 4 is constrained to translate up to the right at 30°. Table 3.3 compares selected grid point displacement, grid point constraint forces, and element forces to those obtained by hand calculation [1].

A solution check (see p. 64) shows an epsilon of 1E-16 and strain energy of 1.6E-9, both of which indicate a desirable solution. Summing reaction forces along the fundamental X coordinate system, accounting for the grid point 4 Y' component in X, yields -1.0 to balance the applied unit load.

Linear Elastic Buckling

The planar frame shown in Figure 3.14 will only buckle if the applied load is to the left such that the element member forces (Table 3.3) are in compression rather than tension. A buckling solution will be performed on this structure using a left-directed force of 1.0E9 N. A refined model with additional grid points is required to display the buckled frame.

Table 3.3   Linear Statics Results

Item[a]	MSC/NASTRAN program	Hand[b] calculation
Displacement (E-9 m)		
u_2	3.198	3.189
v_2	1.007	1.005
u_4	5.504	5.490
Constraint force (N)		
p_1	−0.2216	−0.2209
q_2	0.0475	0.0492
p_3	−0.7757	−0.739
q_3	−0.0521	−0.0517
q_4	0.0053	0.0053
Element axial force (N)		
1	0.12310	NA
2	0.05213	NA
3	0.00515	NA

[a]See Figure 3.1b notation.
[b]Reference [1]. NA = not reported.

Buckling loads and buckled structure deformed shape are accomplished using superelement solutions (see p. 62). A linear static analysis is performed first (SOL 61) applying the desired structure load. A database is generated with internal element forces due to this load. A restart from the database solution uses a buckling solution format (SOL 65). Differential stiffness (p. 27) is calculated based on the initial load, and Eq. (3.8) is solved for λ using an eigenvalue extraction solution. The lowest value of λ is usually desired, and this is the multiplier of the applied load that results in buckling. The buckled structure is then plotted.

An Euler column buckling analysis will give confidence in the method. Table 3.4 shows lowest buckling load for slender columns for three specific end conditions. Figure 3.17a shows the lowest buckling mode for a 5 m steel column, 0.025 m**4 second moment, 2.07E11 N/m**2 elastic modulus, and pinned ends. Ten equal-length BAR elements are used in the model. A

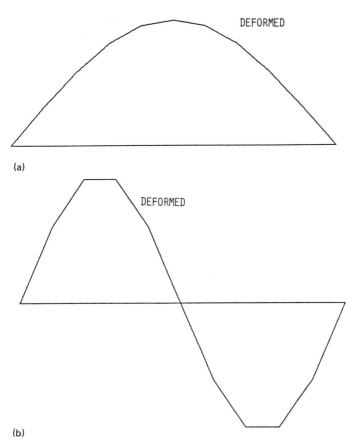

Figure 3.17 Pinned end column, buckling modes. (a) First buckling. (b) Second buckling.

Table 3.4 Euler Column Lowest Critical Loads

End conditions	Buckling load[a]
Pinned-pinned	$E\,I\,\pi^2/L^2$
Pinned-fixed	$2\,E\,I\,\pi^2/L^2$
Fixed-fixed	$4\,E\,I\,\pi^2/L^2$

[a]E = material elastic modulus; I = second moment of inertia; L = column length. Higher buckling loads increase as n^2, where n is an integer.

calculated 2.043E9 N axial compression load agrees with the Table 3.4 pinned-pinned formula. Figure 3.17b shows the second buckled-column deformed shape, which occurs at four times the axial load. This result is in agreement with Euler column buckling theory.

Each of the three steel frame members shown in Figure 3.14 is subdivided into four BAR elements to facilitate displaying buckled deformed shape. Figure 3.18a shows the linear static deformation and Figure 3.18b shows the lowest load buckled shape of the frame structure for the billion N compression load. Figure 3.18b indicates that the pinned beam buckles at a load multiplier (λ) equal to 6.57. Hence, the structure has a minimum buckling load of 6.57 billion N. The fact that Figure 18b shows structure deformation opposite of applied load is insignificant. The figure does indicate that two frame members undergo column buckling from the horizontal applied load.

If gravity loading plays a significant role in structure buckling, the desired value of λ is unity, since it is not possible to arbitrarily increase gravity. This may require buckling solutions with a number of applied loads (added to gravity) in order to achieve a lowest buckling mode multiplier of unity.

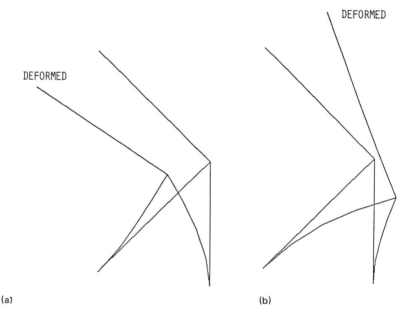

(a)   (b)

Figure 3.18 Example frame, buckling results. (a) Linear statics deformation. (b) Lowest buckling mode, $\lambda = 6.57$.

Real Eigenvalues

Figure 3.19a shows a cantilever beam having 25 cm length, 2.5 cm width, and 0.25 cm thickness. The material is mild steel with a 2.E7 N/cm^2 elastic modulus and 7.84E-5 N sec**2/cm**4 mass density. Several vibrational natural frequencies will be calculated without considering any material damping (see p. 40). A real eigenvalue solution (SOL 3) allows the user to choose one of several eigenvalue extraction techniques [5] with the inverse power method used here.

The cantilever beam is represented by 10 QUAD4 plate elements, and the default lumped-mass method (see p. 29) is compared with the consistent-mass method and theoretical result [16]. A user-controlled parameter selects the consistent-mass method and another user parameter controls root-tracking output to ensure that all lower-frequency modes have been extracted. An alternative generalized dynamic reduction technique extracts all eigenvalues

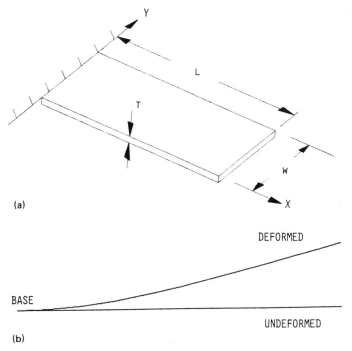

Figure 3.19 Cantilever beam, real eigenvalue example. (a) Physical arrangement. (b) Modal deformation, 34 Hz.

below a user-specified maximum frequency and would be used on a model with more than a few hundred DOF. Single-point constraints specify the built-in boundary condition and remove normal rotational DOF, which have no stiffness for this plate element model.

Natural frequency results are compared in Table 3.5, which shows that the lumped-mass method agrees more closely with the theoretical result than the consistent-mass method. Figure 3.19b shows an edge view of the lowest vibrataion frequency mode shape.

Linear Transient Response

The cantilever beam shown in Figure 3.19 is used to illustrate a modal transient response analysis with a transverse base displacement excitation. This analysis utilizes the enforced-motion technique (see p. 54) and requires that a large mass be added to the model at the excitation point. A user-supplied code selects an enforced displacement, velocity, or acceleration input. A one-cycle triangular-wave transverse base displacement with a half-amplitude of 1 cm and a frequency of either 40 Hz (near resonance) or 100 Hz is input at the cantilever base. Material damping is selected as 5% of critical for the modal transient solutions. Graphic displacement-time plots compare the base and tip transverse displacements.

A MASS element (Table 3.2) of 1000 mass units in the Z displacement DOF is applied at one base grid point, which is rigidly attached to the other base grid point. All other base DOF are constrained. A dynamic load card selects enforced displacement and the material card specifies material damping. This damping forces coupling of the dynamic equations. The lowest four

Table 3.5 Cantilever Natural Frequencies

Mode	Natural frequency (Hz)[a]		
no.	Lumped mass	Consistent mass	Theoretical
1	34.1	34.2	33.1
2	211.0	216.6	207.7
3	584.4	622.1	581.3
4	1132.1	1268.1	1139.6

[a]Theoretical results from Ref. [16].

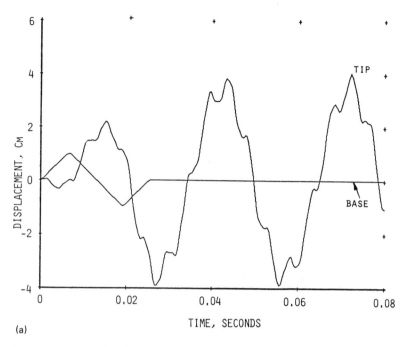

(a)

Figure 3.20 Cantilever beam, enforced transverse displacement. (a) Base excitation: 40 Hz. (b) Base excitation: 100 Hz.

natural frequencies are selected for the modal transient analysis. Figures 3.20a and 3.20b show beam base and tip displacement for the modal transient solution at 40-Hz excitation and 100-Hz excitation, respectively.

Geometric Nonlinear Statics

Small-displacement analysis of the horizontal wire in Figure 3.5a would result in zero transverse stiffness at the load point (p. 29), since ROD element representation has zero rotational stiffness. Hence, a large-displacement (geometric nonlinear) analysis is required to determine equilibrium center wire displacement and internal force in the displaced wire.

The wire length is 500 mm, cross-sectional area 25 square mm, and elastic modulus 2.E5 N/mm^2. A transverse load of 50 kN is applied at the wire center, as shown in Figure 3.21a.

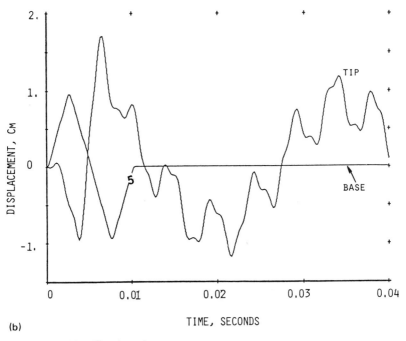

(b)   TIME, SECONDS

Figure 3.20   Continued

A version of geometric nonlinear analysis (Solution 64) may be applied to this problem. The Newton-Raphson method (p. 31) is used with iterations controlled by load subcases. Total load is applied, and each load subcase represents an additional iteration. The user determines convergence by observing that constraint forces, applied by the program to unconstrained grid points, approach zero. Also, the incremental applied work (and resulting internal strain energy)approaches zero. A user convenience allows temporary input of transverse stiffness during early iterations and subsequent removal as the wire deforms. Another user convenience allows automatic iteration until a user-supplied incremental strain energy target is reached. This feature may result in considerable computer use for large models with slow convergence characteristics.

Table 3.6 summarizes program results during successive iterations. The center grid point displacement is rather large in early iterations and converges to -54.5 mm with a 117.4 kN axial force in each rod at equilibrium.

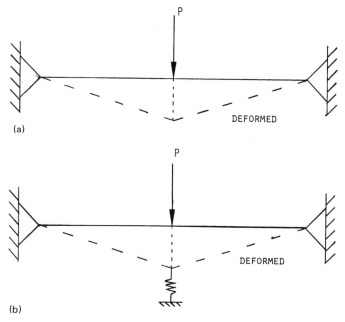

Figure 3.21 Geometric nonlinear statics example, (a) Physical arrangement. (b) Physical arrangement, added transverse stiffness.

Table 3.6 Geometric Nonlinear Analysis Transverse Results (SOL 64)

Iteration	Center displacement (mm)	Incremental external work	Constraint force (N) Ends	Center	Rod Force[a] (N)
1	−2.09E3	5.2E7	20.	5.E4	3.7E7
2	−190.	2.2E8	2.1E6	−4.E6	1.3E6
3	−94.3	2.9E6	1.9E5	−3.4E5	3.4E5
4	−75.5	4.5E5	8.5E4	−1.2E5	2.2E5
5	−54.5	2.3	2.5E4	−1.1E2	1.17E5
Final	−54.5	1.2E-6	2.5E4	−8.0E-2	1.17E5

[a]Axial force in each displaced rod.

Table 3.7 Geometric Nonlinear Analysis Transverse Results (SOL 66)

Applied load (kn)	Center displacement	Constraint force (N)		Rod Force (N)
		1	2	
10	− 31.62	5.0E3	5.0E3	3.98E4
20	− 39.94	1.0E4	1.0E4	6.34E4
30	− 45.81	1.5E4	1.5E4	8.32E4
40	− 50.50	2.0E4	2.0E4	1.01E5
50	− 54.49	2.5E4	2.5E4	1.17E5

Incremental external work and the constraint force on the unconstrained center grid point approach zero at equilibrium.

Solution validity may be easily verified. The deformed rod elements at equilibrium form a 12.3° angle to the horizontal and each rod force has a 25-kN transverse component. Also, the constrained end grid points have 25-kN vertical components for translational force equilibrium. Of course, the wire is well beyond yield for mild steel (200 N/mm^2) with an axial stress of 4696 N/mm^2.

An alternative solution of the taut wire problem shown in Figure 3.21a may be performed using a large displacement parameter in the modern geometric and material nonlinear solution (SOL 66). This solution requires that a negligibly small grounded spring (1 μN/mm) be positioned at the wire center, as shown in Figure 3.21b. The load is then stepped to 50 kN in five load steps. Equilibrium results are shown in Table 3.7, which are in good agreement with Table 3.6 equilibrium results.

Material Nonlinear Statics

A notched tensile-test specimen is shown in Figure 3.22a. The material is mild steel with 2.E5 N/mm^2 elastic modulus and 0.3 Poisson's ratio. Yield stress is 200 N/mm^2, and the stress-strain relationship (see Figures 3.7b) is assumed perfectly plastic with zero slope beyond yield. An inplane tension load is applied at both ends, and material behavior will be observed, especially in the notched region.

Figure 3.22b shows a quarter-model finite element model of the specimen with symmetric boundary conditions on the cut surfaces. A material nonlinear analysis (SOL 66) is applied with the quadrilateral elements in the notched

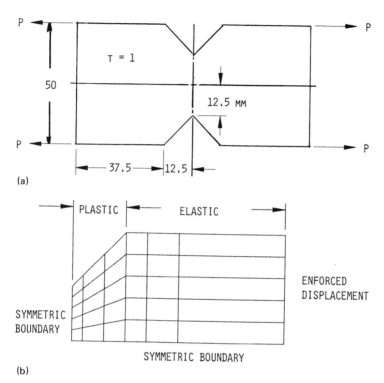

Figure 3.22 Notch tensile specimen. (a) Physical arrangement. (b) One-quarter finite element model.

region assumed nonlinear plastic and the other elements linear elastic. Large displacement effects are ignored. The right end is enforced to the right to simulate load applied by the test machine.

A preliminary elastic analysis shows the element at the bottom of the notch is first to yield at an enforced displacement of 0.0227 mm. Thus, enforced displacements are applied in 0.025-mm increments such that yield occurs on the first increment. Figure 3.23 shows yielding growth in the notch region as a function of enforced displacement. The figure shows complete yielding at the left symmetric boundary at 0.050 mm enforced displacement with greatly increased applied displacements required to yield additional material.

Figure 3.23 results are in good agreement with a more detailed analysis [7] of a similar notched tensile-test specimen.

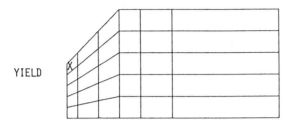

YIELD

(a)

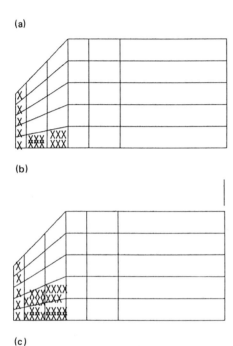

(b)

(c)

Figure 3.23 Notched specimen, yielding results. (a) Enforced 0.025 mm. (b) Enforced 0.05 mm. (c) Enforced 0.3 mm. (d) Enforced 1.0 mm.

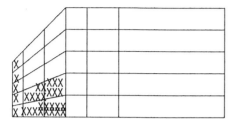

(d)

Figure 3.23 Continued

Material Nonlinear Transient

Geometric and material nonlinear transient analysis (SOL 99) utilizes gap elements, material nonlinear elements, and large displacement effects. Figure 3.24a shows a simple, yet practical, application of this technique. A 485.2-kg mass is suspended on a 500-N/mm elastic spring within a 2426-kg drop container. The container drops 1m and impacts soil with a parabolic force-displacement relationship shown in Figure 3.24b. The objective is to calculate maximum values for suspended mass displacement relative to the container, soil penetration, suspension spring force, and soil impact force. These results are displayed graphically using time history time plots of the several variables.

The suspended mass and container mass are located at two separated grid points using MASS elements, and the suspension spring is modeled by an ELAS element between the two grid points. All DOF are constrained at the two grid points except translation along an axis joining them. A gap element could be used to represent the initial 1000-mm free fall, but it only has linear closing stiffness. Hence, a 10,000-mm-long nonlinear elastic ROD element with unit cross-section area between the container grid point and a constrained grid point simulates free fall and soil resistance force using the following strain-stress pairs for the Figure 3.24b relationship:

Strain	Stress (N/m^2)
0–0.1	0
0.101	− 1000.
0.102	− 4000.
0.103	− 9000.
. . .	. . .
0.115	− 2.25E5

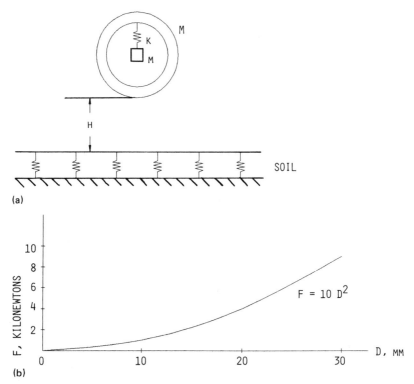

(a)

(b)

Figure 3.24 Drop container, nonlinear transient. (a) Physical arrangement. (b) Soil force-displacement function.

A constant enforced acceleration (see p. 54) acts on the two masses during the entire time period. A total of 55 time steps of 0.01 sec yields a convergent solution. The relative displacement between the suspended and container masses is expressed using an MPC relationship such that a scalar point displacement is equal to the difference in translation displacement of the two mass grid points. No structural or viscous damping is included in the analysis.

Figure 3.25a shows the mass points displacement versus time, and Figure 3.25b shows soil reaction force versus time. Maximum relative displacement between the sprung mass and container is 73 mm, as indicated by a time history plot of the scalar point displacement.

A hand calculation easily shows impact occurs at 0.4515 sec, velocity is 4429.4 mm/sec, and kinetic energy is about 5.9E6 N/mm. These results agree

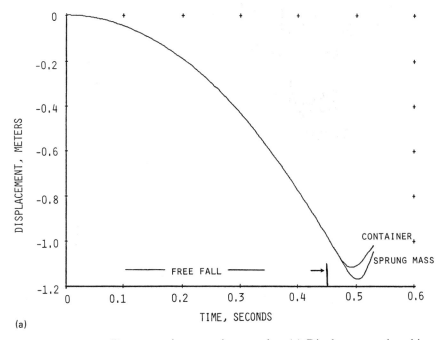

(a)

Figure 3.25 Drop container transient results. (a) Displacement time history. (b) Soil force time history.

with solution results. Other energy balances during the impact phase confirm that solution results are correct. Soil damping could have been included using viscous damping elements available in the program. Of course, soil damping would require addition of a gap element to include damping after impact.

REFERENCES

1. Huston, R. L., and C. E. Passerello, *Finite Element Methods, An Introduction,* Marcel Dekker, Inc., New York, 1984.
2. Cook, R. D., *Concepts and Applications of Finite Element Analysis,* 2nd ed., John Wiley & Sons, Inc., New York, 1981.
3. Schaeffer, H. G., *MSC/NASTRAN Primer,* Schaeffer Analysis, Inc., Mount Vernon, NH, 1982 (contact PDA Engineering, Santa Ana, CA).
4. MacNeal, R. H. (Ed.), *The NASTRAN Theoretical Manual.* MacNeal-Schwendler Corp., Los Angeles, December 1972.

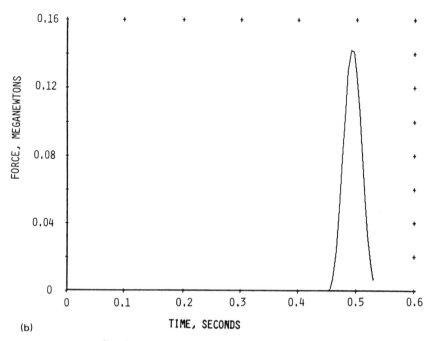

(b)   TIME, SECONDS

Figure 3.25   Continued.

5. Gockel, M. A. (Ed.), *MSC/NASTRAN Dynamic Handbook,* MacNeal-Schwendler Corp., Los Angeles, June 1983.
6. Overbye, V. D., "MSC/NASTRAN Dynamics: Modal or Direct," MSC/NASTRAN Users' Conference Proceedings, Los Angeles, March 20–21, 1986.
7. Lee, S. H. (Ed.), *MSC/NASTRAN Handbook for Nonlinear Analysis,* MacNeal-Schwendler Corp., Los Angeles, 1987.
8. *ABAQUS User's Manual,* Version 4, Hibbitt, Karlsson and Sorenson, Inc., Providence, RI, July 1, 1982.
9. *ANSYS User's Manual,* Revision 4, Swanson Analysis Systems, Inc., Houston, PA, February 1982.
10. *MSC/NASTRAN User's Manual,* Version 64, MacNeal-Schwendler Corp., Los Angeles, 1985.
11. Gockel, M. A. (Ed.), *Handbook for Linear Analysis,* MacNeal-Schwendler Corp., Los Angeles, August 1985.

12. Gockel, M. A. (Ed.), *MSC/NASTRAN Handbook for Superelement Analysis,* MacNeal-Schwendler Corp., Los Angeles, 1981.
13. *PATRAN—User's Guide,* Release 2.0, PDA Engineering, Santa Ana, CA, 1986.
14. Cook, R. D. and W. C. Young, *Advanced Mechanics of Materials,* Macmillan Publishing Co., New York, 1985.
15. Butler, T. G., "Using NASTRAN to Solve Symmetric Structures with Nonsymmetric Loads," MSC/NASTRAN Users Conference Proceedings, Los Angeles, March 18–19, 1982.
16. Seireg, A., *Mechanical Systems Analysis,* International Textbook Co., Scranton, PA, 1969.

4 Thermal Analysis

Vern D. Overbye

The MacNeal-Schwendler Corporation, Milwaukee, Wisconsin

HEAT CONDUCTION INTRODUCTION

General Heat Conduction Equation

The common observation that heat flows spontaneously from high- to low-temperature regions of a structure may be quantified [1–3] as

$$q = -k \frac{\partial T}{\partial n} \tag{4.1}$$

where

q = heat flux

T = temperature

n = a spatial coordinate in the direction of heat flow

k = material thermal conductivity

The negative sign accounts for positive heat flow in the direction of decreasing temperature.

Figure 4.1 shows a parallelepiped of small length dx and unit cross-sectional area. An energy balance on the volume dx may be expressed as

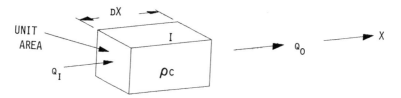

Figure 4.1 One-dimensional energy balance.

$$q_i + Idx = q_0 + \rho C dx \frac{\partial T}{\partial t} \tag{4.2}$$

where

q_i = heat flux in

q_0 = heat flux out

I = specific internal heat generation rate

ρC = thermal capacity

t = time

Expressing heat flux by Eq. (4.1) yields

$$-k \frac{\partial T}{\partial x} + Idx = -k \frac{\partial T}{\partial x} + \left(\frac{\partial}{\partial x}\right)\left(-k \frac{\partial T}{\partial x}\right)dx + \rho C dx \left(\frac{\partial T}{\partial t}\right) \tag{4.3}$$

or

$$\left(\frac{\partial}{\partial x}\right)\left(k \frac{\partial T}{\partial x}\right) + I = \rho C \left(\frac{\partial T}{\partial t}\right) \tag{4.4}$$

Expanding the one-dimensional energy balance in Eq. (4.4) to three dimensions in Cartesian coordinates results in the general heat conduction equation

$$\left(\frac{\partial}{\partial x}\right)\left(k \frac{\partial T}{\partial x}\right) + \left(\frac{\partial}{\partial y}\right)\left(k \frac{\partial T}{\partial y}\right) + \left(\frac{\partial}{\partial z}\right)\left(\frac{k \partial T}{\partial z}\right) + I = \rho C \left(\frac{\partial T}{\partial t}\right) \tag{4.5}$$

One author [3] expresses each of the three left-hand partial derivatives in a more expanded form, as, for example, differentiation with respect to x:

$$\left(\frac{\partial}{\partial x}\right)\left(k_{xx} \frac{\partial T}{\partial x} + k_{xy} \frac{\partial T}{\partial y} + k_{xz} \frac{\partial T}{\partial z}\right) \tag{4.6}$$

This form is used in anisotropic finite element formulations, as shown below.
The thermal conductivity in Eq. (4.5) may be a function of temperature

(temperature-dependent), location (nonhomogeneous), and/or direction (anisotropic). Thermal capacity may also be temperature-dependent and nonhomogeneous. Specific internal heat generation may be nonhomogeneous and/or a function of time.

If thermal conductivity is constant, the Eq. (4.5) relationship yields a term $k/\rho C$ called thermal diffusivity. Material with high thermal diffusivity rapidly propagates the effect of a local temperature change. This is important in estimating the depth at which surface temperature temporal effects are important.

Finite Element Formulation

One-dimensional element

A truss element was used in Chapter 3 (p. 19) to introduce the concept of stiffness in structural analysis. A conductive rod shown in Figure 4.2 may be used to introduce finite element formulation of Eq. (4.1).

Figure 4.2a shows a one-dimensional conductive rod of length L, cross-section area A, and thermal conductivity k. Assuming a linear temperature gradient over the rod length, the heat flow (heat flux times cross-section area) may be expressed as

$$Q_{k1} = \frac{kA}{L}(T_1 - T_2)$$

$$Q_{k2} = \frac{kA}{L}(T_2 - T_1)$$

(4.7)

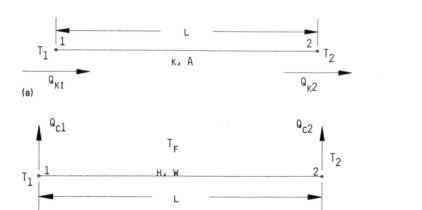

Figure 4.2 One-dimensional thermal rod element. (a) Conduction element. (b) Convection element.

where positive heat flows in the direction of a negative temperature gradient and the subscript k emphasizes conduction. In matrix form this becomes

$$\frac{kA}{L} \begin{bmatrix} 1 & -1 \\ -1 & 1 \end{bmatrix} \begin{Bmatrix} T_1 \\ T_2 \end{Bmatrix} = \begin{Bmatrix} Q_1 \\ Q_2 \end{Bmatrix} \tag{4.8}$$

or

$$[k]\{T\} = \{Q\} \tag{4.9}$$

The analogy between Eq. (4.9) and the truss element force-displacement relationship [Eq. (3.2), Chapter 3] suggests that a structural analysis finite element program may be easily adapted to thermal analysis. In fact, the thermal solution is easier, since temperature at grid points is a scalar rather than a six-displacement-component vector required in structural analysis.

The conduction matrix $[k]$ in Eq. (4.9) contains cross-section area A and rod length L. Hence this matrix expresses a temperature gradient and total heat flow in the element rather than heat flux.

Figure 4.2b shows a surface element (see p. 93) where Q_c represents heat exchange with a surrounding fluid at temperature T_f. The convection surface area is the product of w and L, where w is the rod element perimeter. The heat transfer coefficient h correlates surface heat flow and surface to fluid temperature difference as

$$Q_c = hA_c(T_f - T) \tag{4.10}$$

The matrix form of the surface convection matrix for a rod element is given [6] as

$$[k_c] = \frac{hwL}{6} \begin{bmatrix} 2 & 1 \\ 1 & 2 \end{bmatrix} \tag{4.11}$$

The subscript indicates convection rather than conduction. This surface element convection matrix is combined with the conduction element matrix in thermal analysis finite element programs to form $[K]$. Fluid temperatures are usually formed using scalar points, which represent degrees of freedom (DOF) without physical location in space.

The thermal capacity matrix is given in lumped-mass and consistent-mass form (see p. 29, Chapter 3) as follows:

$$[C_L] = \left(\rho \frac{CAL}{2} \right) \begin{bmatrix} 1 & 0 \\ 0 & 1 \end{bmatrix}$$
$$[C_c] = \left(\rho \frac{CAL}{6} \right) \begin{bmatrix} 2 & 1 \\ 1 & 2 \end{bmatrix} \tag{4.12}$$

Two-dimensional element

A two-dimensional triangular conduction element with linear temperature variation and isotropic material was developed in Chapter 2 (p. 12). This element yields a three row by three column symmetric conduction matrix, which is often referred to as the "element stiffness matrix" [7].

Chapter Preview

Based on the linear temperature gradient finite element concepts given above, the general heat conduction equation will be cast into matrix form [4–6]. Thermal analysis problem types and solution techniques available in commercial finite element programs will be discussed. Structural elements suitable for thermal analysis as well as other elements required in thermal solutions will be described. Boundary conditions, data input and output, solution "goodness numbers," and results verification are explored in some detail. Finally, a number of linear and nonlinear steady-state and transient thermal problems will be formulated, solved, and interpreted using a commercial finite element thermal analysis program.

PROBLEM TYPES

General Conduction Equation, Matrix Form

The analogy between thermal analysis and structural analysis (see p. 88) was early recognized by developers of structural analysis finite element computer programs to extend the application to thermal problems. Thus, letting T replace u in Eq. (3.6), Chapter 3, the general structural analysis equation in matrix form leads directly to the general conduction equation:

$$[B]\{\dot{T}\} + [K]\{T\} = \{P\} + \{N\} \tag{4.13}$$

where

$\{T\}$ = the vector of grid point temperatures

$\{P\}$ = the vector of known time functions of applied heat flow

$\{N\}$ = the vector of temperature-dependent nonlinear heat flow

$[K]$ = the constant heat conduction coefficient matrix

$[B]$ = the constant thermal capacity coefficient matrix

$\{\dot{T}\}$ = the vector of time derivatives of temperature

The term $\{\ddot{u}\}$ in the general structural analysis equation has no significance in thermal analysis.

As in structural analysis, various thermal problems are merely a subset of the general heat conduction Eq. (4.13). Four subclasses will be considered here to account for linear and nonlinear analysis.

Boundary convection and radiation

Convection and radiation are surface phenomena and must be included in Eq. (4.13) to accommodate thermal boundary conditions. Convection on the surface of a conduction element may be expressed as Eq. (4.10). As noted above, the fluid temperature is usually represented by a scalar point, and the convection matrix is combined with the conduction matrix.

Radiation heat flux from a surface depends on surface emissivity and the fourth power of absolute temperature:

$$q = \sigma \epsilon (T + T_a)^4 \tag{4.14}$$

where

q = surface radiative heat flux

σ = Stefan-Boltzmann constant

T_a = a constant to express absolute temperature

ϵ = surface emissivity

Radiation results in a nonlinear finite element model and Eq. (4.13) becomes

$$[B]\{\dot{T}\} + [K]\{T\} + [R]\{T + T_a\}^4 = \{P\} \tag{4.15}$$

where $[R]$ is a radiation matrix (see p. 97).

Linear Analysis

Equation (4.13) has three reduced forms given in Table 4.1 for linear isotropic thermal analysis where thermal conductivity and thermal capacity are constant. The simplified forms are named after famous seventeenth- and eighteenth-century mathematicians. The Laplace and Poisson equations are applicable to steady-state analysis without and with internal heat generation, respectively. The Fourier equation is applicable to transient analysis without internal heat generation, and the general heat conduction Eq. (4.13) is applicable to linear and nonlinear analysis with convection and radiation on structural boundaries.

Linear steady state

Finite element programs generally use the linear static stress approach with a user-specified key such that the analysis is a thermal solution of the Laplace

Table 4.1 Linear Isotropic Thermal Analysis Subsets

Differential formulation[a]	Finite element formulation	Equation name
$\nabla^2 T = 0$	$[K]\{T\} = 0$	Laplace
$\nabla^2 T = I/k$	$[K]\{T\} = \{P\}$	Poisson
$\nabla^2 T - (\rho C/k)(\partial T/\partial t) = 0$	$[B]\{\dot{T}\} + [K]\{T\} = 0$	Fourier
$\nabla^2 T - (\rho C/k)(\partial T/\partial t) = I/k$	$[B]\{\dot{T}\} + [K]\{T\} = \{P\}$	General

[a]$\nabla^2 T = \partial^2 T/\partial x^2 + \partial^2 T/\partial y^2 + \partial^2 T/\partial z^2$.

or Poisson equations. Temperatures are output at grid points, heat flow is shown at constrained grid points, and temperature gradients and heat flux are output for elements. The thermal conduction matrix may be isotropic or anisotropic.

Linear transient

Linear transient solution of Eq. (4.13) without nonlinear terms due to temperature-dependent properties or thermal radiation starts with a specified initial temperature at all grid points. A difference equation based on the Newmark β method (see p. 36, Chapter 3) is expressed as follows:

$$[K]\{\beta\, T_{n+1} + (1 - \beta)\, T_n\} + \left(\frac{1}{\delta t}\right) [B]\{T_{n+1} - T_n\}$$
$$= \{\beta P_{n+1} + (1 - \beta)\, P_n\} \quad (4.16)$$

where

δt = a user-specified time step

β = a user-specified integration stability factor
(between zero and unity, with 0.5 recommended)

n = the nth time step

Finite difference solution methods in linear transient analysis [2] require a restriction on time step size dependent on temperature node mesh size in order to ensure numerical stability of the integration process. The Newmark β method of integration in finite element analysis has no such restrictions for linear transient analysis, and convergence is guaranteed [4] if β equals one-half. Larger values for β are recommended for nonlinear transient analysis if numerical instability is evident.

Nonlinear Analysis

Nonlinear thermal analysis occurs when either conductive elements thermal conductivity or surface elements heat transfer coefficients are temperature dependent, and/or thermal radiation is present. The governing equation is then Eq. (4.13).

Nonlinear steady state

Nonlinear steady state analysis reduces Eq. (4.13) to

$$[K]\{T\} + [R]\{T + T_a\}^4 = \{P\} + \{N\} \tag{4.17}$$

Details for the iterative solution of Eq. (4.17) are found elsewhere [4–6], but in general the Newton-Raphson technique (see p. 31, Chapter 3) is applicable.

An initial estimate of the structure temperature is input such that the conduction and convection matrix may be evaluated using tabular lists of thermal conductivity and/or surface convection coefficient temperature variation. The finite element thermal analysis program iterates to converge to the correct nonlinear temperature distribution. A final conduction matrix update with the final temperature would be desirable before recovery of heat flux in conduction and/or surface elements and heat flow to or from constrained grid points.

Nonlinear transient analysis

Nonlinear transient analysis involves a combination of nonlinear steady-state analysis with Newton-Raphson iteration and linear transient solution technique such as the Newmark β method (p. 91). Solution of models with temperature-dependent thermal capacity is not as common as solution of those involving thermal conductivity and surface convection coefficients. A rather common occurrence of temperature-dependent thermal capacity would concern the heat of fusion in phase change between liquid and solid. Special-purpose techniques are available [5] for these problem types.

THERMAL ANALYSIS PROGRAMS

The engineer who has access to any of the three world-class structural analysis finite element programs, namely, ABAQUS, ANSYS, and MSC/NASTRAN (see p. 40, Chapter 3), also has access to thermal analysis [8–10]. Also, the two electromagnetic finite element programs described in Chapter 5, namely, MSC/MAGNETIC and MSC/MAGNUM, have steady-state thermal capability.

All thermal analysis elements, loads, constraints, and example problems presented in this chapter utilize the MSC/NASTRAN finite element program owing to the author's familiarity with and access to this program.

COORDINATE SYSTEMS

Structural Analysis Similarity

Several coordinate systems described in Chapter 3 (p. 41) are used in finite element thermal analysis. The local displacement coordinate system and displaced element coordinate system for large displacement effects has no meaning in thermal analysis. The element coordinate system is used to display up to three orthogonal heat fluxes in elements. The material coordinate system is used to orient material axes and output element heat flux for three-dimensional elements.

ELEMENT TYPES

Conduction Elements

Table 3.1 in Chapter 3 showed that ABAQUS and ANSYS have special finite elements required exclusively for thermal analysis. All of the MSC/NASTRAN elements tabulated in Table 3.2 in Chapter 3 are suitable for heat conduction, except the SHEAR and TRIAX elements. The axisymmetric element alternatives are a four-grid-point TRAPRG and three-grid-point TRIARG element.

The one-dimensional ROD element is adequate for thermal analysis, but the BAR, BEND, and BEAM elements are included as a user convenience. The TRIA and QUAD elements use only membrane properties for thermal analysis. The conduction material axis may be oriented as shown in Figure 3.10b in Chapter 3.

Surface Elements

The ABAQUS and ANSYS finite element programs allow surface convection and/or radiation heat transfer directly into conduction element surfaces. However, the MSC/NASTRAN program requires that surface elements be used. Table 4.2 summarizes these elements.

Table 4.2 Selected MSC/NASTRAN Thermal Surface Elements

HBDY type	Number of grid points	Surface application	Normal definition
POINT	1	Line end	User
LINE	2	Line periphery	User
REV	2	Axisymmetric	Connectivity
AREA3	3	Triangular	Connectivity
AREA4	4	Quadrilateral	Connectivity

Element connectivity is important in radiation analysis as the element active surface has an outward-directed normal determined by element connectivity and the right-hand rule.

Element Properties

Conduction elements reference an isotropic (MAT4) or anisotropic (MAT5) material card, which specifies element thermal conductivity and thermal capacity. Surface elements reference isotropic material cards that specify the convection coefficient.

MATERIAL PROPERTIES

Thermal conductivity of materials may vary in direction such that heat flux differs for a given temperature gradient. The three major finite element thermal analysis programs accommodate linear and nonlinear (temperature-dependent) isotropic and anisotropic conductive materials. Chapter 3 (p. 51) describes terms used in definition of constitutive relationships that are quite similar in thermal analysis.

Linear Isotropic Material

Isotropic materials are defined by a single value for thermal conductivity and thermal capacity. Linear material properties are temperature independent and may be rotated with respect to element axis (see Figure 3.10b) for two-dimensional elements. Three-dimensional element material properties may be specified in the element coordinate system, the fundamental coordinate system, or an arbitrary local coordinate system. Element thermal heat flux is output in the material coordinate system.

Linear Anisotropic Material

A three-dimensional anisotropic material is completely defined with a three row by three column symmetric material matrix defined as

$$[K] = \begin{bmatrix} k_{xx} & k_{xy} & k_{xz} \\ k_{xy} & k_{yy} & k_{yz} \\ k_{xz} & k_{yz} & k_{zz} \end{bmatrix} \tag{4.18}$$

where x, y, and z are the three orthogonal material axes. A two-dimensional element is defined by three conductivity coefficients. These coefficients are temperature independent for linear thermal analysis.

Nonlinear Temperature-Dependent Material

Isotropic and anisotropic material properties may be formed using tables of temperature-thermal conductivity pairs for steady-state and transient analysis. The MSC/NASTRAN finite element program does not accept temperature-dependent thermal capacity in either nonlinear steady-state or nonlinear transient analysis.

LOADS

Steady-State Loads

Thermal analysis models may be loaded by constraining grid points to specified temperature, specifying specific internal heat generation within conduction elements, or specifying heat flux (including directed radiation) on surface elements. These thermal loads within conduction elements or on surface elements are resolved into connecting grid points.

Grid point loads

Grid points and scalar points (a DOF not physically located in space) may be constrained to an arbitrary temperature (including zero) using single-point constraints (see p. 97) or an enforced temperature (SPCD) that resembles an applied grid point load rather than a constraint. A grid point constraint uses DOF one and a scalar point constraint uses DOF zero.

Conduction element loads

Internal heat generation is applied to the volume of conduction elements by specifying an internal heat generation rate per unit volume. These thermal powers are then lumped into connecting grid points. For example, a rectangular hexahedron element (HEXA) with eight corner grid points would lump one-eighth of the element power into each grid point.

Surface element loads

Surface elements shown in Table 4.2 may be supplied with nondirected heat flux or directed heat flux. Nondirected heat flux times element area is lumped into connecting grid points. Directed heat flux such as incident radiation from the sun is multiplied by element area, surface absorptivity, and the cosine of the angle between the incoming radiation and the element normal. Radiation from behind the element is ignored and all reflected radiation is considered lost.

Transient Loads

Any of the steady-state thermal loads into conduction or surface elements may be made time dependent by specifying time variation in tabular and/or harmonic form [4–6]. The structural analysis enforced motion technique (see p. 54, Chapter 3) has an equivalent in thermal analysis. An excellent grounded thermal conductor is formed using an ELAS element (Table 3.2, Chapter 3), which is attached to the grid or scalar point to be enforced by a prescribed time history. A multiplier equal to the ELAS conductivity is applied to the grid point temperature-time tabular relationship. Thus, the grid point exhibits the desired time history regardless of surrounding structure thermal conductivity.

CONSTRAINTS AND BOUNDARY CONDITIONS

Thermal analysis generally involves heat sources and heat sinks represented by constrained temperatures at grid points and/or scalar points. Single-point constraints (SPC, see p. 55, Chapter 3) are used for this purpose to enforce temperatures to a specified value, including zero. In addition, several areas of a structure may be enforced to a common (undefined) temperature relationship between grid points using multipoint constraints (MPC, see p. 55, Chapter 3). Adiabatic surfaces have zero heat flow, which is the natural boundary condition for unconstrained grid points on structure surfaces.

This section discusses thermal applications of SPCs, MPCs, natural and other boundary conditions, and partial models of symmetric structures with symmetric thermal loads.

Single-Point Constraints

Thermal analysis has 1 DOF per grid or scalar point, and grid points are constrained with DOF component one and scalar points with DOF component zero. Linear steady-state analysis involves decomposition of the conduction matrix into triangular factors for each change in constraint set. Hence, changing specified temperatures in several load cases using SPC sets would involve several decompositions of the constrained conduction matrix. An alternative is to use an SPCD set, which forces constrained grid points to resemble thermal loads, and only one conduction matrix decomposition is required during an analysis.

Multipoint Constraints

Multipoint constraints (see p. 55 Chapter 3) may be used for an arbitrary temperature relationship between several grid or scalar points for linear analysis. Nonlinear analysis is restricted to an equivalencing between two grid or scalar points.

Thermal Boundary Conditions

Thermal analysis conduction models may have adiabatic, specified temperature, specified heat flux, radiation heat flux, and/or convection heat flux at boundaries. Structure boundaries are adiabatic (see p. 124, Chapter 5) if surface grid points are unconstrained and no surface elements are present. Boundary surface temperature is specified using SPC and/or MPC relationships. Specified input surface heat flux is an applied thermal load. Radiation and convection surface heat flux is considered when surface elements are present. The convection heat transfer coefficient matrix is part of the conduction matrix, and radiation properties such as view factors and surface emissivity and absorptivity make up a radiation matrix in Eq. (4.17).

The radiation view factor is a geometric parameter that accounts for the fraction of radiation leaving a surface finite element i that is intercepted by surface finite element j. The radiation view factors are calculated in an MSC/ NASTRAN VIEW module, and these results are used along with surface element area and emissivity to form $[R]$.

Symmetry

The rubber-sheathed current-carrying wire shown in Figure 2.1 (Chapter 2) contains quadrilateral geometric symmetry about a vertical and horizontal centerline. A quarter-model would be adequate with adiabatic boundary conditions on the cut surfaces if the surface convection heat transfer coefficients and ambient temperature are symmetric on the sheath surface and if the specific internal heat generation rate is symmetric about the two centerlines. If this is not the case, a full model would be required to input spatial distribution of boundary conditions and internal heat generation.

Therefore, thermal analysis problems are not as amenable to dihedral or rotational cyclical symmetry as structural problems (see p. 56, Chapter 3). Also, the use of periodic boundary conditions by means of MPC relationships is more common in magnetic field modeling (see p. 125), Chapter 5) than in thermal analysis. However, if an adiabatic boundary can be located in a geometrically symmetric structure, the use of a partial model should be employed.

RESULTS AND VERIFICATION

This section summarizes some of the checks the engineer should make to ensure that the thermal analysis finite element model is satisfactory, surface elements are present and properly oriented for radiation, the solution is numerically satisfactory, and the results are reasonable.

Structural Analysis Model Similarity

Usually most structural analysis elements may be used directly in a thermal analysis. Hence, all the preliminary considerations described in pages 62–63 (Chapter 3) are directly applicable to the thermal model.

Surface Element Plotting

Surface elements must be present to control surface heat transfer by convection and radiation. Surface elements require no particular orientation for convection, but a radiation application requires that the active surface normals face each other in order to effect radiation heat transfer. These surface normals are determined by element connectivity for axisymmetric and two-dimensional surface elements.

Structural plotting of surface elements verifies their presence. Some finite element model generation programs [11] will plot surface element normals to verify that active surfaces are facing each other in radiation analysis.

Solution Verification

Solution "goodness" numbers were shown for structural analysis (p. 64, Chapter 3) wherein the ratio of virtual work done by calculated displacements to work done by applied loads is formed. This same epsilon value is calculated for thermal analysis as follows:

$$\{\delta P\} = \{P\} - [K]\{T\}$$

$$\epsilon = \frac{\{T\}^T \{\delta P\}}{\{T\}^T \{P\}}$$

(4.19)

where $\{P\}$ equals applied linear thermal load.

This ratio is indicative of solution accuracy and should approach zero for well-conditioned finite element models.

Results Verification

Thermal analysis results may be verified by making certain that the second law of thermodynamics is not violated in that heat does not spontaneously flow to a location of higher temperature. Thus, a structure being cooled by convection to a surrounding fluid should have no temperature below ambient. However, such results do appear occasionally, especially in nonlinear steady-state analysis. Possible sources of such erroneous results may be

Warped elements
Incorrect fluid temperature specification on surface elements
Program error

Warped elements may be subdivided or rearranged and surface element ambient temperature specification corrected. Often a nonisoparametric element is more stable for thermal analysis. Also, removing radiation surface elements near sharp corners on conduction elements may correct program errors.

An equilibrium steady-state thermal analysis solution requires that heat flow from heat sources be equal to heat flow to heat sinks. The program will output element volumes and areas such that estimates of surface heat flux and internal generation may be made. Also single-point constraint heat flows are useful in performing an overall energy balance.

Temperature contour plots are invaluable in visually examining temperature profiles in a structure. These data may be output as isotherms or color contour plots showing temperature profile. Unexpected temperature gradients may be used to detect modeling errors or incorrect specification of boundary conditions.

THERMAL ANALYSIS EXAMPLES

This section contains several examples of finite element applications to linear and nonlinear steady-state and transient thermal analysis. Additional example problems are available [5,12].

Linear Steady State

Linear steady-state thermal analysis requires solution of the Laplace or Poisson equation (Table 4.1) with constant thermal conductivity. The Laplace solution is independent of thermal conductivity, whereas the Poisson solution depends on thermal conductivity and internal heat generation rate. Boundary conditions must be linear and may involve specified temperature, adiabatic areas, and/ or convection heat transfer on surfaces.

Laplace solution

Figure 4.3 shows a 1-m^2 region 0.1 m thick with an arbitrary thermal conductivity of 1.0 W/m°C. The finite element model consists of TRIA elements (see Table 3.2, Chapter 3) with grid point identification at element corners and element identification in element centers. Table 4.3 shows constrained

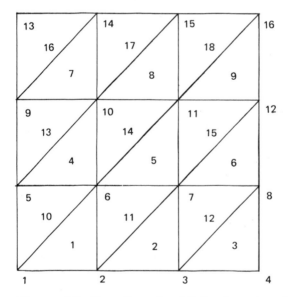

Figure 4.3   Two-dimensional finite element model.

Table 4.3 Laplace Solution Temperatures

Grid point[a] identification	Temperature (°C)	
	Constrained	Calculated
1, 13	5.	
5, 9	10.	
2–4, 8, 12, 14–16	0.	
6, 10		3.75
7, 11		1.25

[a]See Figure 4.3.

surfaces temperatures on all edges with four interior grid point temperatures to be calculated.

Boundary grid points are specified using single-point constraints in this small model. Larger models with several different temperature profiles would be efficiently solved using an SPCD card described on page 95. A linear static structural analysis is used to solve linear steady-state thermal problems with a flag to indicate heat transfer analysis. Results for the four interior grid point temperatures are also tabulated in Table 4.3 and are in good agreement with a hand calculation [7] using the linear temperature element derived in Chapter 2.

Poisson solution

The Poisson equation for linear steady-state thermal analysis accommodates internal heat generation and the solution depends on thermal conductivity. An example finite element model described in Chapter 2 will be solved using a quarter-model and a half-model to satisfy two different types of imposed boundary conditions.

Figure 4.4a shows a quarter-model of the rubber-sheath copper wire shown in Figures 2.1 and 2.2 (Chapter 2). The 8 by 14 mm wire is coated with rubber for an outside dimension of 12 by 20 mm. Thermal conductivity is 0.375 W/mm°C and 0.2 mW/mm°C for the copper and rubber, respectively. Uniform wire resistance heating results in 1 mW/mm^3 internal heat generation. A uniform heat transfer coefficient is 0.009 mW/mm^2°C and ambient temperature is 0°C. The objective is to determine maximum wire temperature and heat loss per mm wire length.

TRIA and QUAD plate elements 1 mm thick model the quarter-model. Cut surfaces are adiabatic to represent the other three quadrants. The exposed

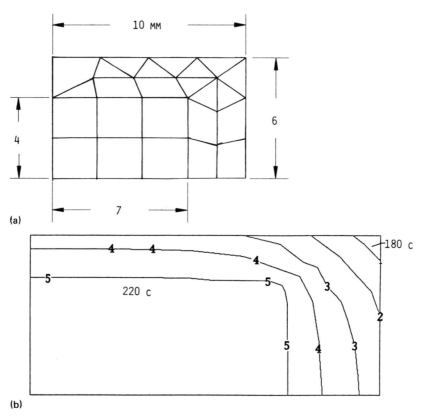

Figure 4.4 Sheathed wire, quadrilateral symmetry. (a) Finite element model. (b) Isotherms, free convection cooling.

surfaces are modeled with line HBDY elements (see Table 4.2) with a 1-mm width. The ambient temperature is modeled with a scalar point constrained to 0°C. Internal heat generation is applied using QVOL cards for the wire elements.

Figure 4.4b shows isotherms that differ by 10°C from 180° to 220°C. Maximum wire temperature is 223.8°C and the rubber surface is about 200°C, except at the corner. Summing constraint "forces" yields a wire loss of 28 mW/mm length for the quarter-model, which is in agreement with the total internal heat generation input in the wire.

Figure 4.5a shows a half-model of the sheathed copper wire with the left vertical boundary adiabatic to represent a symmetric boundary condition. The

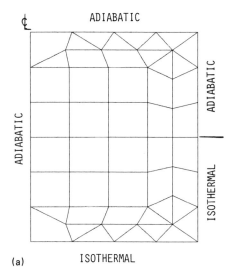

(a)

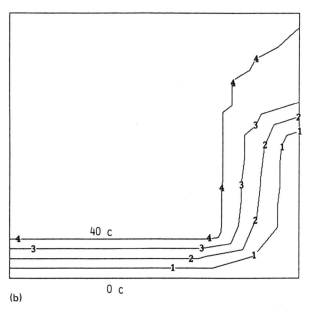

(b)

Figure 4.5 Sheathed wire, half-model. (a) Finite element model and boundary conditions. (b) Isotherms.

bottom rubber surface and lower half of the right vertical surface are assumed in contact with a high-convection coefficient fluid at 0°. The top surface and upper right surface are assumed adiabatic to represent negligible air cooling compared to the fluid. Thermal conductivities and internal heat generation are the same as in the first example. The objective is to determine maximum wire temperature and temperature profile.

Rubber surface grid points in contact with the fluid are constrained to 0°C. Figure 4.5b shows solution isotherms that differ by 10°C from 10° to 40°C. The upper rubber surface is nearly equal to wire temperature. Maximum temperature is 44.75°C. Again, single-point constraint "forces" on the lower rubber isothermal surface equal total internal heat generation.

Linear Transient

Figure 4.6 shows a large 120-mm-thick plate initially at 38°C suddenly immersed in a 260°C fluid with an infinite surface convection coefficient. The plate thermal conductivity is 70 mW/mm°C and the thermal capacity is 11 mW sec/mm^3°C. Plate temperature profile up to 180 sec is required.

This is a one-dimensional linear transient thermal analysis, and a model of half-thickness is adequate with an adiabatic boundary at the plate centerline. The initial temperature is specified as 38°C and the surface is constrained at 260°C for all time. An XY plot of temperature versus time for the surface, quarter-depth, and plate centerline displays temperature history.

Six 10-mm ROD elements (see p. 93) model the half-thickness. Thermal conductivity and thermal capacity are specified on a homogeneous material card. A highly conductive ELAS element is connected between the surface grid point and ground. A multiplier equal to the ELAS conductivity inputs the constant 260°C surface temperature during the transient period (see p. 96). A 10-sec time step results in adequate output data to plot temperature versus time.

Figure 4.6b shows temperature history on the plate surface, quarter-depth, and plate centerline. These results are in good agreement with a hand calculation using a finite difference model [2].

Nonlinear Steady State

Temperature-dependent conductivity

The large plate shown in Figure 4.7a has a thickness of 160 mm and a 260°C temperature on the left face and 40°C on the right face. Thermal conductivity varies linearly between 70 and 7 mW/mm °C at 260° and 40°C, respectively. Steady-state temperature distribution through the plate is required.

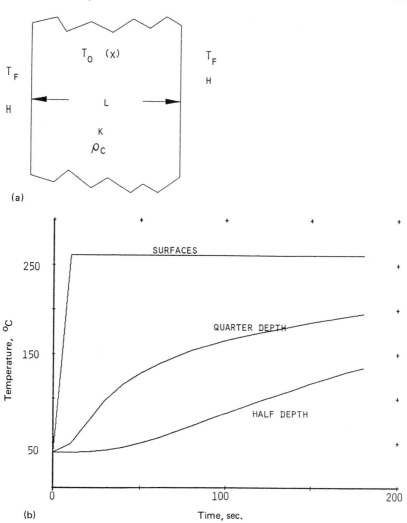

Figure 4.6 Linear transient analysis. (a) Physical arrangement. (b) Temperature-time variation.

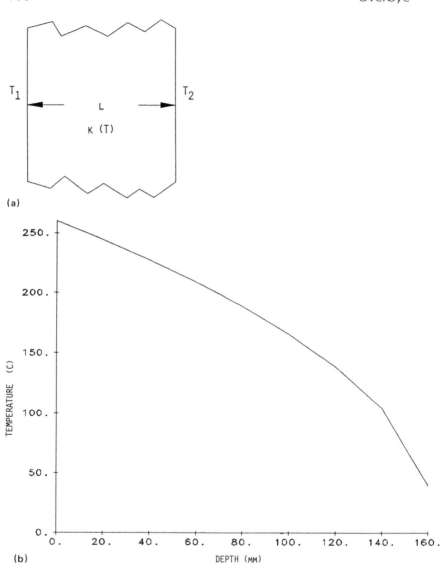

Figure 4.7 Nonlinear steady-state conduction example. (a) Physical arrangement. (b) Temperature profile.

The plate is modeled with eight 20-mm ROD elements. Thermal conductivity is made temperature-dependent using a MATT card, which points to a temperature-conductivity table. Surface temperatures are fixed at 260° and 40°C for left and right end grid points, respectively.

Nonlinear steady-state analysis requires an initial temperature estimate, and the conduction matrix is assembled using this initial temperature profile. Hence, a second thermal analysis is required using calculated temperatures as initial temperature estimate in order to recover heat flux in the ROD elements. Initial temperature estimates are best specified high to enhance modified Newton-Raphson convergence (see p. 92), so 260°C is used in this example.

Figure 4.7b shows wall temperature profile, and a second analysis with this temperature estimate yields the same heat flux in all elements. This indicates that correct thermal conductivity at element temperature is obtained from the table.

Radiation

Radiation heat transfer depends on fourth power of absolute temperature and requires a nonlinear steady-state solution. Geometric view factors and surface emissivity are important in calculating radiation exchange between black and/ or gray bodies. Also, a "collector" radiation exchange element allows finite element model simplification by accounting for stray radiation, as will be shown in this example.

Figure 4.8a shows two 0.5-m^2 plates separated 0.5 m with temperatures of 1000° and 500°C and surface emissivities of 0.2 and 0.5, respectively. The objective is to determine heat lost by each plate to black-body surroundings at 27°C. Only facing surfaces are considered active.

Two AREA4 HBDY elements (see Table 4.2) represent the surfaces with grid point connectivity directing the active-surface normal, and property cards specify emissivity and absorptivity. Execution of the VIEW module (see p. 97) using 64 subelements calculates a 0.28784 view factor between the two surfaces. This geometric result agrees with published data [2] and indicates that 28.8% of emitted radiation is received by the opposite surface. A two row by two column symmetric radiation matrix would contain the view-factor–surface-area product.

The surroundings are represented by a "collector" HBDY element with a view factor of 0.71216 with respect to each surface and added to the radiation matrix, as shown in Figure 4.8b. Only the lower triangular factor is input to the program. A POINT HBDY element with an area of 0.71216 m^2 may be

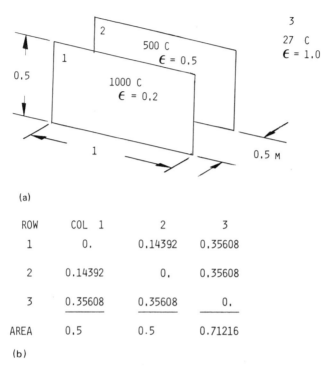

(a)

```
ROW      COL  1         2          3
 1        0.         0.14392    0.35608

 2       0.14392      0.        0.35608

 3       0.35608    0.35608      0.

AREA     0.5         0.5        0.71216
```

(b)

Figure 4.8 Radiation between plates and surroundings. (a) Physical arrangement. (b) Radiation matrix.

located at an arbitrary grid point and emissivity and absorptivity specified on a material card.

The nine grid points used for this model are constrained at appropriate temperatures, which are also used for initial temperature estimate. Absolute temperature conversion and the Stefan-Boltzmann constant are input on parameter cards. The model is completely constrained, so an additional grid point must be connected to an arbitrary grid point with a ROD element to allow calculation of one temperature.

Calculated results show 14.4 kW and 2.6 kW loss from the high- and low-temperature surface, respectively, and 17 kW is absorbed by the surroundings. This is in good agreement with a hand calculation [13]. A second run is unnecessary since the intial temperature is correct.

It is recommended by the author that "collector" surface elements be used for stray radiation in all radiation problems. This example could be used to simulate radiation loss to space by constraining the surroundings temperature at $-273°C$.

Nonlinear Transient

Figure 4.9a again shows the Figure 4.7a model with identical thermal conductivity. Thermal capacity is 7 mW sec/mm³°C and initial temperature is 260°C. At zero time the right face is reduced to 40°C and maintained at that temperature. The objective is to determine plate temperature profile to steady state and compare with the earlier nonlinear steady-state result.

ELAS elements maintain plate surface temperatures as in the linear transient model (see p. 96). A nonlinear transient solution is used that allows only thermal conductivity and not thermal capacity to be temperature dependent. Thirty 100-sec time steps are adequate to achieve steady-state temperature profile in agreement with Figure 4.7b. Temperature history at the plate surfaces, quarter-depth, and centerline are shown in Figure 4.9b.

Two runs are unnecessary in nonlinear transient analysis because the conduction matrix is updated at each user-requested output time. Thus heat flux is correct at all times. This frequent matrix update will cause large models to require considerable computer time compared with linear transient analysis.

REFERENCES

1. Carslaw, H. S., and J. C. Jaeger, *Conduction of Heat in Solids,* 2nd ed., Oxford University Press, Cambridge, England, 1959.
2. Kreith, F., *Principles of Heat Transfer,* 3rd ed., Intext Press, Inc., New York, 1973.
3. Cook, R. D., *Concepts and Applications of Finite Element Analysis,* 2nd ed., John Wiley & Sons, Inc., New York, 1981.
4. MacNeal, R. H. (Ed.), *The NASTRAN Theoretical Manual,* MacNeal-Schwendler Corp., Los Angeles, December 1972.
5. Boothe, W. H. (Ed.), *MSC/NASTRAN Thermal Analysis,* MacNeal-Schwendler Corp., Los Angeles, 1987.
6. Lee, H. W., *NASTRAN Thermal Analyzer Manual,* Vol. I, No. X3227616, Goddard Space Flight Center, Greenbelt, MD, December 1975.
7. Huston, R. L., and C. E. Passerello, *Finite Element Methods, An Introduction,* Marcel Dekker, Inc., New York, 1984.

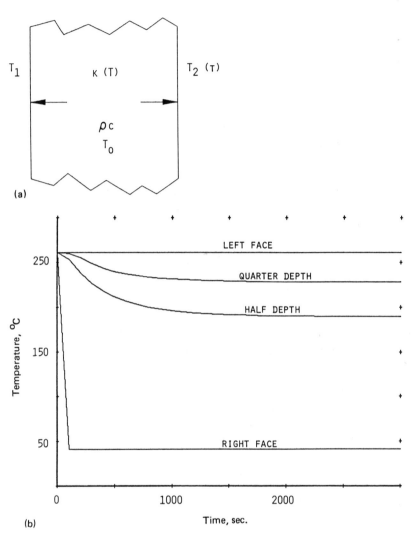

(a)

(b)

Figure 4.9 Nonlinear transient conductor example. (a) Physical arrangement. (b) Temperature-time variation.

8. *ABAQUS User's Manual,* Version 4, Hibbitt, Karlsson and Sorenson, Inc., Providence, RI, July 1, 1982.
9. *ANSYS User's Manual,* Revision 4, Swanson Analysis Systems, Inc., Houston, PA, February 1982.
10. *MSC/NASTRAN User's Manual,* Version 64, MacNeal-Schwendler Corp., Los Angeles, 1985.
11. *PATRAN User's Guide,* Release 2.0, PDA Engineering, Santa Ana, CA, 1986.
12. Jackson, C. E., *NASTRAN Thermal Analyzer Sample Problem Library,* Vol. II, No. X3227617, Goddard Space Flight Center, Greenbelt, MD, December 1975.
13. Holman, J. P., *Heat Transfer,* 4th ed., McGraw-Hill Book Co., Inc., New York, 1976.

5 Electromagnetic Analysis

John R. Brauer

The MacNeal-Schwendler Corporation, Milwaukee, Wisconsin

PROBLEM TYPES

Electrical engineers often design devices in which electric, magnetic, or elec-tromagnetic fields are important. The calculation of these fields and related device performance parameters, such as electrical power input, power output, losses, efficiency, and impedance, can be achieved using finite element anal-ysis. Usually metric (MKS) units are used for these parameters, as will be done throughout this chapter.

Electrostatic Fields

Electrostatic fields obey Poisson's equation, which in two dimensions was discussed on page 12. That section also derived the finite element matrix expression, in which potential is the grid point quantity. Here the potential is called the electric scalar potential ϕ measured in volts, and Poisson's equation [Eq. (2.5)] becomes for nonconducting regions in two dimensions

$$\frac{\partial}{\partial x}\epsilon\frac{\partial\phi}{\partial x}+\frac{\partial}{\partial y}\epsilon\frac{\partial\phi}{\partial y}=-\rho \tag{5.1}$$

where ϵ is dielectric constant or permittivity of the materials and ρ is charge per unit volume. In three dimensions Eq. (5.1) becomes

$$\frac{\partial}{\partial x}\epsilon\frac{\partial\phi}{\partial x}+\frac{\partial}{\partial y}\epsilon\frac{\partial\phi}{\partial y}+\frac{\partial}{\partial z}\epsilon\frac{\partial\phi}{\partial z}=-\rho \qquad (5.2)$$

In either case the electric field $\bar{E}$ is a vector defined by

$$\bar{E}=\frac{-\partial\phi}{\partial x}\frac{-\partial\phi}{\partial y}\frac{-\partial\phi}{\partial z}=-\nabla\phi \qquad (5.3)$$

In many problems the charge density ρ is zero, and ϕ is known to be zero volts (ground) in some areas and to be a certain high voltage in other areas. Examples include high-voltage insulators and switchgear.

The energy functional for electrostatic problems is

$$F = \int_v \left[\frac{\epsilon E^{**}2}{2}\right] dv - \int_v \left[\frac{\phi\rho}{2}\right] dv \qquad (5.4)$$

where the first term is the stored electrostatic energy and the second term is the input electrostatic energy.

Steady Current Flow

Also obeying Poisson's equation are steady electrical currents flowing in good conductors. In three dimensions the equation is

$$\frac{\partial}{\partial x}\sigma\frac{\partial\phi}{\partial x}+\frac{\partial}{\partial y}\sigma\frac{\partial\phi}{\partial y}+\frac{\partial}{\partial z}\sigma\frac{\partial\phi}{\partial z}=0 \qquad (5.5)$$

where σ is the electrical conductivity of the various materials.

As always the electric field $\bar{E}$ is given by Eq. (5.3). The current density $\bar{J}$ is given by the vector form of Ohm's law:

$$\bar{J} = \sigma\bar{E} \qquad (5.6)$$

Power loss per unit volume p is given by

$$p = \frac{J^{**}2}{\sigma} \qquad (5.7)$$

Knowledge of J and p over a region containing wires and busbars can determine electrical efficiency and heating.

Electrodynamic

The above two problem types involve static electric fields. Time-varying electric field problems in conducting dielectric materials are here called electrodynamic problems. The magnetic field is assumed to have negligible effect on voltage and current.

$\sigma = \frac{I}{V} = \frac{U \cdot S}{V}$ $\quad \epsilon = \frac{q}{V}$

stationary ϕ or hi-freq (insulator)

$\epsilon / \sigma = \frac{q}{I} = \frac{\text{coul}}{\text{amp}} \cdot s$

$\sigma/\epsilon = \frac{S}{\omega}$

electrostatics electrodynamics steady-state current

dielectric relaxation time

relaxed ϕ or low-freq (conductor)

$\frac{\sigma}{\epsilon} \sim \omega\phi$

$\frac{d}{dt} \angle E = \text{ } (\neq 0)$

for SiPCS
$\omega \sim 10^9$ Hz
$\epsilon \sim 10^{-10}$ F/m
$\omega\epsilon \sim 10^{-1}$ F/ms
$\sigma = 10$ kΩ·cm
$= 10^{-4}$ U/cm
$\sigma \sim 10^{-2}$ U/m

In a general electrodynamic problem the materials have nonzero values of electrical conductivity σ and permittivity ϵ. An example would be a lossy capacitor energized by an alternating voltage of given frequency. The frequency f multiplied by 2π is the angular frequency ω. If the conductivities σ are all much less than ω times the dielectric constants ϵ, then an electrodynamic solution is not necessary; instead electrostatic solutions at each instant of time can be used. If instead σ is much greater than ω times ϵ, the E fields at each instant can be found by steady current flow solutions at each instant "snapshot" of time. If ϵ and ω times ϵ are both the same order of magnitude, then the E field may vary in phase angle over the region analyzed and an electrodynamic solution is required. The stiffness matrix for an electrodynamic solution is complex rather than the real matrix of static problems.

Linear Magnetostatic

The magnetic field $\overline{B}$ is a vector quantity, as is the electric field $\overline{E}$. $\overline{B}$ is also called magnetic flux density and governs the performance of a wide range of magnetic and electromagnetic devices. Examples include electric motors, generators, and actuators.

$\overline{B}$ is also related by material permeability μ to the magnetic field intensity vector $\overline{H}$ by

$$\overline{B} = \mu\overline{H} \tag{5.8}$$

Such a relation is linear and occurs when μ is constant.

Just as $\overline{E}$ is related to a potential called voltage ϕ, B is related to a magnetic potential $\overline{A}$. However, unlike ϕ, which is a scalar, $\overline{A}$ is a vector called magnetic vector potential. The relationship defining $\overline{A}$ is

$$\overline{B} = \nabla x \overline{A} \qquad \overline{B} = \nabla \times \overline{A} \tag{5.9}$$

In two-dimensional linear problems where $\overline{B}$ is invariant with z, it can be shown that $\overline{A}$ obeys Poisson's equation in x and y:

$$\frac{\partial}{\partial x}\frac{1}{\mu}\frac{\partial A}{\partial x} + \frac{\partial}{\partial y}\frac{1}{\mu}\frac{\partial A}{\partial y} = -J \qquad \frac{\partial^2 A_z}{\partial x^2} = -\mu J_z \tag{5.10}$$

where A is solely the z component of magnetic vector potential, and J is also in the z direction.

In three-dimensional problems or in axisymmetric problems Poisson's equation is not obeyed by $\overline{A}$ or $\overline{B}$. In three-dimensional problems all three components of $\overline{A}$ must be included in the calculations. In axisymmetric problems only the component of $\overline{A}$ normal to the rz plane need be calculated.

The energy functional for linear magnetostatic problems is

$$F = \int_v \left[\left(\frac{B^{**}2}{2\mu} \right) - \left(\frac{\overline{J} \cdot \overline{A}}{2} \right) \right] dv \qquad (5.11)$$

where the first term is the stored magnetic energy and the second term is the input electrical energy. The derivation of the finite element matrix equation in two dimensions is similar to that described on page 12.

If $\overline{J}$ is known in Eq. (5.11) everywhere at any given instant of time, then time-varying magnetic field problems can also be solved using these magnetostatic techniques. This "snapshot" approach is similar to the one mentioned in the preceding section for electric field problems.

Nonlinear Magnetostatic

Many, if not most, magnetic devices are made of steel, which has a linear B versus H relation or constant μ only at B values lower than used in most designs. Therefore, it is essential to be able to analyze nonlinear materials.

The energy functional for nonlinear magnetostatic problems is [1]

$$F = \int_v \left(\int_0^B \overline{H} \cdot d\overline{B} \right) dv - \int_v \left(\int_0^A \overline{J} \cdot d\overline{A} \right) dv \qquad (5.12)$$

where the first term is the energy stored in saturable (nonlinear) or linear (constant permeability) materials, and the second term is the input electrical energy in linear or nonlinear problems.

If the permeability μ is not constant, then the stiffness matrix depends on the magnitude of B (and J). An iterative procedure is developed by expanding the partial derivative of F with respect to A into a multidimensional Taylor's series

$$\left| \frac{\partial F}{\partial A_i} \right|_{A + \delta A} = \left. \frac{\partial F}{\partial A_i} \right|_A - \left. \frac{\partial^2 F}{\partial A_i \partial A_j} \right|_{A + \delta A} \delta A_j + \cdots \qquad (5.13)$$

where i and j are integers varying from 1 to the number of grid points N. Setting Eq. (5.13) to zero gives the matrix equation

$$\delta A_j = - \left| \frac{\partial^2 F}{\partial A_i \partial A_j} \right|_{A + \delta A}^{-1} \left. \frac{\partial F}{\partial A_i} \right|_A \qquad (5.14)$$

This equation is the basis of the Newton (also called Newton-Raphson) iterative process used to solve for A in a saturable magnetic device. The Jacobian matrix in Eq. (5.14) is first estimated from an initial solution using approximate material permeabilities. Then Eq. (5.14) is solved repeatedly until the correction δA_j is negligibly small. In each solution of Eq. (5.14) both the Jacobian matrix and the residual vector in its right-hand side are reevaluated based on the latest A values, enabling rapid convergence to the correct saturable potentials A throughout the device.

The exact expressions for the Jacobian matrix and residual vector are derived elsewhere for two-dimensional problems [2]. The technique requires knowledge of reluctivity $\nu\,(= 1/\mu)$ and of $\partial\nu/\partial(B^2)$ in each nonlinear material. These parameters are obtainable from the *B–H* curves supplied as input material data, as will be described on page 121.

Magnetodynamic

Magnetic fields that vary with time can induce currents in conducting materials. If the induced currents are unknown but large enough to significantly alter the magnetic field from the static field of the preceding two sections, then the problem is called magnetodynamic. In such problems one usually wants to calculate the induced or eddy currents as well as the total resulting magnetic field.

From Eq. (5.6), induced current density $\bar{J}_I$ is given by

$$\bar{J}_I = \sigma\bar{E}_I \tag{5.15}$$

where the induced electric field $\bar{E}_I$ is given by Faraday's law:

$$\nabla \times \bar{E}_I = -\frac{\partial\bar{B}}{\partial t} \tag{5.16}$$

Substituting Eq. (5.9) gives

$$\bar{E}_I = -\frac{\partial\bar{A}}{\partial t} \tag{5.17}$$

The total electric field is then the sum of Eq. (5.3) and Eq. (5.17):

$$\bar{E} = -\nabla\phi - \frac{\partial\bar{A}}{\partial t} \tag{5.18}$$

Now voltage is the integral of electric field $\bar{E}$(in volts/meter) along a line 1:

$$V = \int \bar{E} \cdot d\bar{l} \tag{5.19}$$

Thus voltage is related to both electric potential ϕ and magnetic vector potential $\bar{A}$:

$$V = \phi - \int \frac{\partial \overline{A}}{\partial t} \cdot d\overline{l} \qquad (5.20)$$

To calculate induced currents and related magnetic and electric fields, the losses due to the induced currents must be included in the functional F. For magnetodynamic problems with sinusoidal excitation and constant permeability (linear) materials, F becomes [3]

$$F = \int_{v} \left(\frac{B^2}{2\mu} - \overline{J} \cdot \overline{A} + i\,\omega\,\frac{1}{2}\,\sigma\,\overline{A}^2 \right) dv \qquad (5.21)$$

where $\overline{J}$ is the applied current density of angular frequency ω, μ is permeability, and σ is conductivity. The resulting magnetic fields and currents are all sinusoidal with various phase angles and may be expressed as complex numbers called phasors. The matrix equation resulting from minimizing F of Eq. (5.21) is also complex, leading to longer solution times than for linear magnetostatic problems with the same number of grid points. Because phasors are all of the same frequency and nonlinearities introduce multiple frequencies, the materials must be linear in Eq. (5.21).

Electromagnetic

When coupled electric and magnetic fields exist, they are said to constitute an electromagnetic field. Such coupled fields cannot be calculated separately because of their strong interaction. Examples of coupled-field problems are many high-frequency devices such as antennas, waveguides, and resonators with transverse electric (TE) or transverse magnetic (TM) fields. Coupled electromagnetic problems require time-varying excitation.

The functional for electromagnetic problems must include all the energy terms of the electric and magnetic functionals of the previous six sections, plus several additional coupling terms. Thus, in general both $\overline{A}$ and ϕ are unknowns, although at high frequencies ϕ is often negligibly small from Eq. (5.18). The program MSC/MAGNUM [4] solves coupled electromagnetic problems with or without ϕ.

The simultaneous existence of $\overline{E}$ and its stored energy and $\overline{B}$ and its stored energy allows resonance. That is, at certain frequencies large amounts of energy oscillate back and forth between electric and magnetic fields. The magnetic fields have an equivalent inductance L. The electric fields have an equivalent capacitance C. Thus, an electromagnetic finite element model can be equivalent to an LC circuit, which resonates at select frequencies.

ELEMENT TYPES

Electromagnetic finite elements are of the types presented in Chapter 2. Two-dimensional and axisymmetric electromagnetic elements include triangles and quadrilaterals. Such elements for magnetic problems have vector potential $\overline{A}$, applied currents, and induced currents all assumed directed out of the plane of the elements. Three-dimensional electromagnetic finite elements include hexahedrons, pentahedrons, and tetrahedrons, all allowing all three components of $\overline{A}$ and currents.

Unlike structural finite element modeling, air and vacuum must be modeled with electromagnetic finite elements. If significant magnetic or electric fields may exist in a region of air or vacuum, then that region must be broken up into finite elements. The air (or vacuum) finite elements must have some grid points common to those of adjacent finite elements made of steel or other material.

In many magnetic devices very small air gaps exist between large steel poles. Very long and thin finite elements may be used in such cases. Accurate field calculations are obtainable if the grid points accurately define the air gap spacing, and if each air gap finite element is expected to have an approximately uniform $\overline{B}$.

In many cases there are significant $\overline{B}$ or $\overline{E}$ fields extending fairly far outside the device to be modeled. Rather than modeling the surrounding air or vacuum with many layers of finite elements and grid points, open-boundary or infinite elements can be used in the program MSC/MAGNETIC [5]. These elements are discussed further on page 127.

MATERIAL PROPERTIES

Three properties are needed to describe materials in electromagnetic problems. These properties are all bulk quantities determined by the molecular or domain structure of the material and are valid as long as all the material dimensions are much greater than the size of a molecule or of a magnetic domain. These properties are valid for solid materials, but may not completely characterize materials that are liquids, gases, semiconductors, or plasmas.

The three electromagnetic material properties are permeability, permittivity (or dielectric constant), and electrical conductivity. All three may vary with temperature or frequency or with other physical parameters. (not time)

Permittivity ϵ of Eq. (5.1) is defined by

$$\epsilon = \frac{\overline{D} \quad \text{C/m}^2}{\overline{E} \quad \text{V/m}} \tag{5.22}$$

$C/_{n^2}$ $V/_m$

where $\overline{D}$ is called the electric flux density and and $\overline{E}$ is the electric field from page 114. If $\overline{D}$ and $\overline{E}$ are in the same direction, then the permittivity is isotropic and is a scalar. If $\overline{D}$ and $\overline{E}$ are in different directions, then ϵ is a tensor.

If the ratio of Eq. (5.22) is a constant, then the permittivity is a constant. Most materials have constant permittivity, because the relationship between $\overline{E}$ and $\overline{D}$ is linear. For example, air or vacuum has permittivity $\epsilon_o = 8.854\mathrm{E}-12$ farads/m. Commonly other permittivities are expressed relative to air; thus, relative permittivity equals ϵ/ϵ_o.

There exist materials where $\overline{D}$ is a nonlinear function of $\overline{E}$. However, nonlinear permittivity is not nearly as common in engineering problems as is nonlinear magnetic permeability.

Permeability μ is defined by Eq. (5.8). Air or vacuum has permeability $\mu_0 = 12.57\mathrm{E}-7$ henries/m. Usually other permeabilities are given relative to air; that is, relative permeability equals μ/μ_o. Most materials have a relative

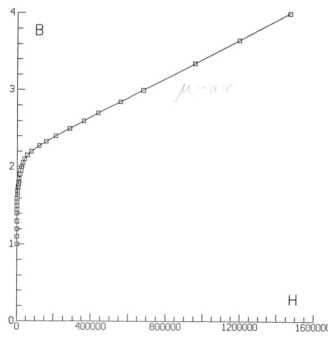

Figure 5.1 *B–H* curve for typical SAE 1010 steel, commonly used for small electric motors and other magnetic devices.

Table 5.1   B and H Points on Figure 5.1

B (T)	(amps/m)
0.0	0.0
1.0	295
1.1	340
1.2	400
1.3	500
1.4	600
1.45	790
1.50	960
1.55	1250
1.60	1620
1.643	2387
1.674	3183
1.700	4000
1.733	5570
1.749	6366
1.778	7958
1.805	9540
1.845	11940
1.904	15915
1.955	19900
2.000	24000
2.050	31000
2.100	40000
2.150	55600
2.200	76000
2.275	119366
2.330	158000
2.400	210000
2.500	282000
2.600	358000
2.700	437800
2.850	557000
3.000	680000
3.354	955000
3.655	1.1936E6
4.000	1.4682E6

permeability of one. Relative permeability μ_r of a typical steel is several thousand if B is in the linear region of the steel $B-H$ curve.

Figure 5.1 and Table 5.1 describe the $B-H$ relationship for a typical steel used in electric motors. Note that, as described on page 116, the relationship is highly nonlinear for B above roughly 1.5 teslas (T). The steel saturates at about 1.5 T, carrying proportionately much less B as H increases. Note, however, that at high B the slope of the $B-H$ curve approaches the permeability of air. The slope is never zero, but varies from several thousand to one. For rapid convergence of the Newton algorithm of Eq. (5.14) the slope of the $B-H$ data points must be smooth. As seen in Figure 5.1, the slope of the $B-H$ curve varies tremendously. It is a good idea to plot the slope versus B or B^2 to verify the $B-H$ data.

Electrical conductivity σ is defined by Eq. (5.6). Air, vacuum, and other insulating materials have a conductivity of zero, and they conduct no current density J unless E is high enough to cause arcing. The most common conductor, copper, has $\sigma = 5.8E7$ S/m at room temperature. In most materials σ varies somewhat with temperature, but is independent of $\overline{E}$. Exceptions are doped semiconductors, in which σ varies not only with the magnitude but with the polarity (direction) of $\overline{E}$. Materials that do not have constant electrical conductivity are not analyzed in this book.

Steel and iron have a conductivity in the range of about 1.E6–1.E7 S/m. If B changes with time, then to reduce the induced losses of Eq. (5.15) steel is very often laminated. The steel laminations (sheets) are separated by a very thin layer of air and/or oxide, which effectively lowers the losses. Also, lowering the induced currents ensures that they are not large enough to reduce B.

EXCITATIONS

There are two ways of exciting or creating a magnetic field: electrical currents and permanent magnets.

Electric current density $\overline{J}$ creates a magnetic field $\overline{B}$ according to

$$\nabla \times \frac{1}{\mu} \overline{B} = \overline{J} + \frac{\partial \overline{D}}{\partial t} \tag{5.23}$$

where $\overline{D} = \epsilon \overline{E}$, and its time derivative is called displacement current. If displacement current is zero, Eq. (5.23) leads to Poisson's equation [Eq. (5.10)] for the two dimensional case only.

Equation (5.23) with negligible displacement current can be integrated over a surface S to give Ampere's law:

$$\int \overline{H} \cdot d\overline{l} = \int \overline{J} \cdot d\overline{s} \qquad (5.24)$$

The units of the right-hand side are amperes. Besides simply increasing current, the right-hand side can be increased also by increasing the number of conductors N each carrying a current I, because Eq. (5.24) gives

$$\int \overline{H} \cdot d\overline{l} = N I \qquad (5.25)$$

Thus H is proportional to the product of N and I. To keep I low, many magnetic devices have coils of wire containing many conductors N.

The other way of creating $\overline{B}$ is with permanent magnetization. Figure 5.2 shows a typical $B\text{–}H$ curve for a permanent magnet. Note that at $H = 0$ the curve has nonzero B, called residual flux density B_r. At $B = 0$ the curve has $H = -H_c$, where H_c is called coercive force. The B created by the permanent magnet within itself is some value between 0 and B_r, depending on its shape and surroundings. The distribution of $\overline{B}$ within and outside permanent magnets is calculated by finite element analysis on page 151.

Sometimes the source of magnetic field $\overline{B}$ is outside the device analyzed, and the currents or permanent magnetization that create $\overline{B}$ are unknown. An

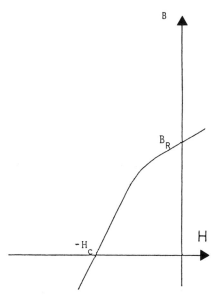

Figure 5.2 *B–H* curve for a typical permanent magnet material.

example is the earth's magnetic field. Such an applied $\overline{B}$ can be enforced by specifying vector potential $\overline{A}$ to obtain $\overline{B}$ on the device boundaries according to Eq. (5.9).

Electric fields $\overline{E}$ are in general related to both $\overline{A}$ and ϕ according to Eq. (5.18), which simplifies to Eq. (5.3) for static problems independent of time. Thus one way of exciting $\overline{E}$ is to specify $\overline{A}$ and/or ϕ.

Another excitation of $\overline{E}$ is with electric charges such as charge density ρ, according to Poisson's equation [Eq. (5.1)]. The charges may be at grid points, expressed in coulombs, or in finite elements, expressed usually in coulombs per cubic meter.

BOUNDARY CONDITIONS

Electric and magnetic fields can be confined or unconfined. For example, a current carrying coil in air has magnetic fields extending to infinity. Surrounding the coil by steel essentially confines the field $\overline{B}$ within the outer boundary of the steel. Thus, in addition to boundary conditions that confine electromagnetic fields within a finite element model, infinite boundary conditions are also useful. Usually constraints are needed only on the boundaries of the finite element model, not within it.

The boundary conditions are on magnetic vector potential $\overline{A}$ and on electric scalar potential ϕ. Constraints on $\overline{A}$ govern magnetic field problems and will be discussed first, followed by constraints on ϕ.

Two-Dimensional A

In two-dimensional planar problems a line of constant magnetic vector potential A is called a magnetic flux line. Flux lines form a pattern similar to that obtained by sprinkling long, thin iron filings or compass needles on the plane. For most electric machines $\overline{B}$ flows in the plane of the steel laminations and the flux is assumed confined to the steel outer boundary. Flux lines along such a boundary (not crossing it) are enforced by setting $A = 0$ along the boundary.

Many electric machines have identical poles, or even identical half-poles. The number of grid points can be greatly reduced if the mesh need only contain one pole or one-half pole. For example, a mesh containing one-half pole often has flux lines parallel to one radial boundary and perpendicular to the other radial boundary. Absence of any constraint on an exterior grid point can be shown to cause the flux lines to be perpendicular to the finite element mesh boundary. This perpendicularity is called the natural boundary condition.

In any electric machine having identical poles, each pole boundary has periodic boundary conditions. For rotary planar machines periodic boundary conditions are expressed in polar (r, θ) coordinates as

$$A (r, \theta_o + p) = -A (r, \theta_o) \tag{5.26}$$

where θ_o is the angle of one radial boundary and p is the pole pitch angle. This multipoint constraint (MPC) is often called a NEGA boundary condition. Figure 5.3 has flux line patterns that show that only one pole pitch need be modeled in a machine with identical poles. Figure 5.3 also shows that the pole pitch modeled may be a piece of any shape, as long as at all radii the radial boundaries are one pole pitch apart. Thus, the rotor may be rotated by up to one pole pitch from the stator in a one-pole-pitch model representing the entire machine. If the geometry requires modeling two poles, then the As on the boundary are set equal (POSA) with no negative sign. Generally an

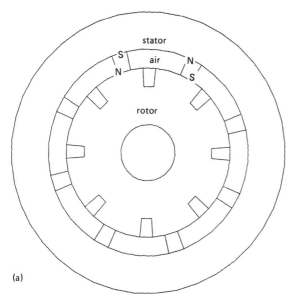

(a)

Figure 5.3   Modeling one pole pitch of an eight-pole motor using periodic boundary conditions on A. (a) Entire motor. (b) One possible division into eight identical pieces of a puzzle. (c) Two-dimensional finite element model of one piece containing one pole pitch. (d) Flux plot of entire motor. An identical picture can be obtained by assembling eight copies of a flux plot of a one-pole-pitch model with NEGA boundary conditions.

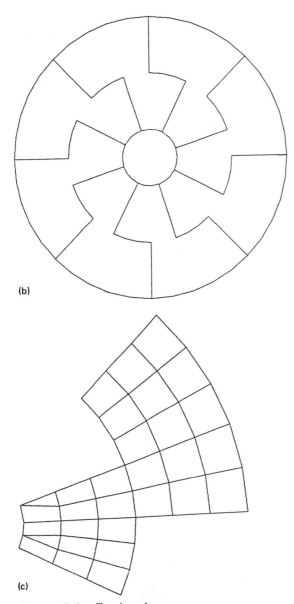

(b)

(c)

Figure 5.3 Continued

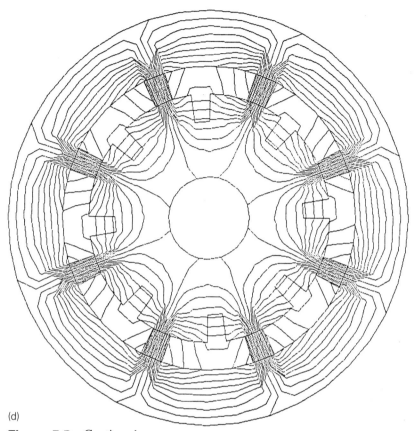

(d)

Figure 5.3 Continued

odd number of poles requires NEGA and an even number requires POSA boundary conditions.

When the amount of steel in a finite element model is small or zero, infinite boundary conditions are used. It is impractical to model large amounts of air or vacuum far away from the device analyzed. The infinite boundary conditions are specified by surrounding the device by one layer of "infinite" or open-boundary elements, which are especially formulated to extend the model far out into space. Typical open-boundary finite elements extend the boundary to a distance exceeding $1. \times 10^{100}$ light-years.

Axisymmetric A

In axisymmetric problems a magnetic flux line is a line of constant radius (from the axis) times A. A and J are in the peripheral ϕ direction out of the plane of the model.

Because magnetic flux cannot cross the axis of symmetry, A must be constrained to zero along the axis. If the device is surrounded by steel, then A is also set to zero on its outer boundaries. Occasionally an axisymmetric device has periodic symmetry and MPCs are used. If no constraints are on a boundary, then the natural boundary condition enforces the flux lines to be normal to the boundary.

Infinite or open-boundary finite elements are available in axisymmetric problems to model fields extending to infinity. They have been shown to give highly accurate results for problems such as circular coils in free space [6].

Three-Dimensional $\overline{A}$

The existence of three vector components of $\overline{A}$ in three-dimensional magnetic and electromagnetic problems makes boundary conditions more complicated than in two-dimensional or axisymmetric problems. The three components of $\overline{A}$ may be A_x, A_y, and A_z, or in cylindrical devices such as generators and motors $\overline{A}$ is conveniently expressed in the three cylindrical components A_r, A_θ, A_z.

While plots of contours of constant A in two-dimensional problems are flux line plots, in three dimensions flux line plots are not rigorously defined. It is difficult to envision sprinkling iron filings throughout three dimensions. Nevertheless, contours of constant total magnitude of $\overline{A}$ appear to have meaning analogous to planar flux plots.

Boundary conditions of $\overline{A}$ in three dimensions are governed by the curl of $\overline{A}$ in Eq. (5.9) expanded, for example, in xyz coordinates:

$$\overline{B} = \left(\frac{\partial A_z}{\partial y} - \frac{\partial A_y}{\partial z}\right)\overline{u}_x + \left(\frac{\partial A_x}{\partial z} - \frac{\partial A_z}{\partial x}\right)\overline{u}_y + \left(\frac{\partial A_y}{\partial x} - \frac{\partial A_x}{\partial y}\right)\overline{u}_z \quad (5.27)$$

Thus one way of enforcing $B_z = 0$ is to set $A_x = A_y = 0$. Carrying this technique further, $\overline{B}$ can be prevented from crossing a boundary surface by setting the tangential components of $\overline{A}$ to zero.

Another boundary condition is $\overline{B}$ normal. This is the natural boundary condition, occuring when no constraints on $\overline{A}$ are enforced on the boundary.

Periodic boundary conditions are also common, especially in machines such as motors and generators. These periodic conditions involve A_x, A_y, A_z

or A_r, A_θ, A_z, depending on whether the periodicity is translational or rotational. The appropriate constraints are usually multipoint constraints (MPCs) between like components of A. For example, boundaries one pole pitch apart on a rotary machine might have A_r the negative of each other, A_θ the negative of each other, and A_z the same. The best way to determine the correct boundary conditions is to make a coarse finite element model of the entire device and observe the relationship obeyed by $\overline{A}$ at grid points one pole pitch (or other distance) apart. Then these constraints are applied to a fine finite element model of that portion of the device.

Infinite elements do not yet exist for three-dimensional magnetic problems. Instead, where fields are not confined, it is necessary to place finite elements out to a large distance from the source. At the outer boundary surface $\overline{B}$ is very small, and thus the appropriate components of $\overline{A}$ are constrained to zero.

Electric Potential

Electric potential ϕ of Eq. (5.1) and Eq. (5.2) is often constrained on inner and boundary grid points. It is constrained wherever the static voltage is known. Examples include high-voltage electrodes and ground planes in DC and 60-Hz electrical apparatus.

An outer boundary without constraints obeys the natural boundary condition. The natural condition is that the contours of constant potential ϕ are normal to the boundary. Thus $\overline{E}$, the gradient of ϕ, is tangential to the boundary.

Periodic boundary conditions occasionally are appropriate for ϕ. An example is a high-voltage electric machine, in which the voltage reverses sign from one pole to the next. MPCs are appropriate for ϕ, with either negative or positive signs depending on whether an odd number or even number of poles is modeled.

Infinite boundary conditions are quite common for ϕ. Unless the electrical device analyzed is surrounded by electrical conductors, ϕ gradually decreases as one moves away from the apparatus analyzed. Thus infinite finite elements are used as described on page 127.

RESULTS AND VERIFICATION

The basic results calculated by an electromagnetic finite element program are numerical values of the magnetic and/or electric potential at every grid point in the finite element model. Although the potentials are of interest to the engineer, other, more important results are obtainable. Many of these results are needed to verify that the device analyzed can perform satisfactorily. The results discussed below are usually of great interest to the design engineer. They help verify the finite element model and point the way to better product design.

Contour Plots of Potential

The best way to quickly understand the results of finite element analysis is to obtain a contour of the calculated potential. Figure 5.3 is an example.

The contour plot shows, first, whether the desired boundary conditions have been properly applied. For example, if many contour lines go to a boundary, then that boundary has not been constrained to zero.

In the case of two-dimensional magnetic field problems, such as Figure 5.3, the density of the A contours is proportional to flux density $\overline{B}$, as described on page 124. The direction of the lines is the direction of $\overline{B}$. Thus the contour plot is a flux plot that tells the designer where steel should be added to avoid saturation and where it can be removed to save cost and space.

For axisymmetric magnetic problems, as described on page 128, the flux plot is a plot of contours of constant radius times A. The direction of the flux lines is the direction of $\overline{B}$, but the magnitude of $\overline{B}$ is not proportional to the inverse of the spacing between flux lines.

For three-dimensional magnetic and electric field problems contour plots may be made in various surfaces. Interactive plotting is especially helpful because many surfaces must often be examined.

Contour plots of electric potential ϕ are useful in understanding electrostatic fields E. From Eq. (5.3), $\overline{E}$ is the negative gradient of ϕ, and thus $\overline{E}$ is highest where the ϕ contours are closest. The direction of $\overline{E}$ is normal to the ϕ contours.

$\overline{B}$ or $\overline{E}$

The magnetic field $\overline{B}$ and electric field $\overline{E}$ are calculated and printed out for each finite element. The field value at each element centroid is obtained from the adjacent grid point $\overline{A}$ and ϕ values using Eq. (5.9) and Eq. (5.18), respectively.

A particularly effective way for the design engineer to learn the distribution of $\overline{B}$ or $\overline{E}$ is a color display. Usually the highest field magnitudes are displayed in red and the lowest in blue. A color spectrum with a scale of values enables the engineer to quickly determine the field magnitude anywhere on the display within one or two significant figures. If a color display terminal is not available, a less effective monochromatic display with numerically labeled field strength contours may suffice. The goal is to aid the designer in shaping the materials for best performance. Often the optimum shape of magnetic devices is obtained by adding steel where B is above 1.5 T and subtracting steel where B is below 1.5 T, obtaining a device with B everywhere about 1.5 T, just below saturation.

Measurements of $\overline{B}$ in air gaps can be made using Hall effect probes to confirm the calculations. Measurements of $\overline{B}$ within steel or magnets are usually impossible, making finite element calculation extremely valuable.

Flux

Magnetic flux is very useful in obtaining voltage and inductance. From $\overline{A}$ the flux flowing between any two points is easily obtained. The definition of magnetic flux is

$$\psi = \int \overline{B} \cdot d\overline{s} \tag{5.28}$$

Substituting the definition of $\overline{A}$ from Eq. (5.9) gives

$$\psi = \int \nabla \times \overline{A} \cdot d\overline{s} \tag{5.29}$$

By Stokes' identity the surface integral may be replaced by a closed-line integral around the surface:

$$\psi = \oint \overline{A} \cdot d\overline{l} \tag{5.30}$$

Thus for two-dimensional problems the flux between any grid points 1 and 2 is simply

$$\psi_{12} = (A_1 - A_2)d \tag{5.31}$$

where d is depth (stack) into the page.

For axisymmetric magnetic problems recall that $\overline{A}$ is in the peripheral direction. Thus the length of a flux line is $2\pi r$, where r is the radius. Hence the flux of Eq. (5.31) becomes for axisymmetric problems

$$\psi_{12} = 2\pi \, (r_1 A_1 - r_2 A_2) \tag{5.32}$$

In three-dimensional magnetic problems Eq. (5.30) cannot in general be simplified. The contour integral may be evaluated over any three-dimensional closed loop in space. An interactive feature of the MSC/MAGNUM three-dimensional program computes the flux through any contour described by the user. The user simply enters the grid point numbers in order around the contour.

Flux in electric field problems may be defined as

$$\psi = \int \overline{D} \cdot d\overline{s} \tag{5.33}$$

Because electric potential is a scalar voltage, Eq. (5.33) cannot be replaced by a line integral. However, electric flux is not usually needed by the design engineer.

L, C, R

The design engineer usually needs to calculate the inductance L, capacitance C, and resistance R of the device analyzed. All these circuit parameters can be obtained by finite element analysis.

From the magnetic vector potential A the inductance L is calculable. L for linear or for saturable problems is calculated using its definition:

$$L = \frac{\lambda}{I} \tag{5.34}$$

where I is current and flux linkage λ is defined as

$$\lambda = N\psi \tag{5.35}$$

where N is the number of turns of the coil. If I and λ are for the same coil, then L is called self-inductance. If they are for different coils, then L is the mutual inductance between the two coils.

Using the equations for flux ψ of the preceding section, L can be calculated by the finite element program. For example, for two-dimensional problems it can be shown [1] that the self-inductance per unit depth of a coil is

$$\frac{L}{d} = \frac{J}{3I^2} \sum_{n=1}^{N_j} \left(S_n \sum_{k=1}^{3} A_k \right) \tag{5.36}$$

where d is depth, J is current density, N_j is the number of elements making up one side (one slot) of the coil, and S_n is the area of each triangular finite element.

The capacitance C is calculable from the electric field E. The energy stored in E is

$$W_E = \frac{1}{2} \int_v \epsilon |\overline{E}|^2 \, dv \tag{5.37}$$

which is calculated by summing over all finite elements in the programs MSC/MAGNETIC and MSC/MAGNUM. Capacitance obeys the relation

$$W_E = \frac{1}{2} CV^2 \tag{5.38}$$

where V is the voltage difference applied across the capacitor electrodes. Assuming the same V is used as a constraint in the finite element analysis, one obtains

$$C = 2 \frac{W_E}{V^2} \tag{5.39}$$

Resistance R is known without finite element analysis in many cases, such as for coils at low frequency made of uniform cross-section wire. If the cross-section is not uniform, then a steady current flow solution, as described on page 114, is often necessary. The finite element program may print out the total power loss:

$$P = \int_v \frac{1}{\sigma} J^2 \, dv \tag{5.40}$$

Another expression for P is from basic circuits:

$$P = VI \tag{5.41}$$

where V is the voltage difference, here assumed to constrain the finite element model.

Using Ohm's Law

$$R = \frac{V}{I} \tag{5.42}$$

Combining the preceeding three equations gives

$$R = \int_v \frac{1}{\sigma} J^2 \frac{dv}{I^2} \tag{5.43}$$

Impedance

From basic electric circuit techniques, impedance Z is related to R, L, and C connected in series by

$$Z = R + j \omega L - j \frac{1}{(\omega C)} \tag{5.44}$$

where $\omega = 2\pi$ times frequency f. L and C determine the imaginary component of Z, called the reactance X.

At low frequencies the R, L, and C equations of the preceding section apply. However, as mentioned on page 117 for magnetodynamic problems, above a certain frequency (which may be only 1 Hz in some cases) the magnetic field is affected by eddy currents. Thus R, L, and C may vary with frequency. It is then best to calculate impedance directly if possible.

Magnetodynamic solutions described on page 117 obtain the first two terms of the impedance of Eq. (5.44). This Z is calculable using the definition

$$Z = \frac{V}{I} \tag{5.45}$$

where V and I are complex quantities. If there are no skin effects (see p. 151) within the conductor carrying the current I, then V is given by Faraday's law as

$$V = -N \frac{\partial \psi}{\partial t} \tag{5.46}$$

Using Eqs. (5.30), (5.45), and (5.46) gives

$$Z = -N \frac{\partial}{\partial t} \int \overline{A} \cdot \frac{d\overline{\ell}}{I} \tag{5.47}$$

Assuming I and A are sinusoidal of angular frequency ω gives

$$Z = j \omega N \int \overline{A} \cdot \frac{d\overline{\ell}}{I} \tag{5.48}$$

For example, for axisymmetric problems

$$Z = \frac{i\omega A_{ave} N}{I_s} 2\pi r_o \tag{5.49}$$

where r_o is the average radius (distance to the axis of symmetry) of the source current, A_{ave} is the average phasor magnetic potential over the source current region, and N is the number of conductors carrying I_S in each conductor.

Losses

Calculation of power losses of electromagnetic devices is important for two reasons. First, losses determine the efficiency of a device. Second, power losses generate heat which must be removed to avoid burnout or fire.

The resistance R of the preceding two sections has an associated power loss $I^2 R$. Thus, for DC currents Eq. (5.40) gives the power loss. In most cases the DC resistance R of wire is known from wire tables, and finite elements are not required to calculate this loss.

In high-frequency AC problems eddy current losses occur in all finite elements with conductivity σ greater than zero. The loss per unit volume is J^2/σ, where J is the total of induced and applied current densities. The summation of these losses over the entire model should equal the input power I^2R, where R is the real part of the impedance Z of Eq. (5.48) or (5.49).

In AC magnetic devices a loss called core loss occurs in the laminated steel "core." As described on page 121, eddy currents in laminations are usually not large enough to significantly affect the magnetic field. Core loss consists of eddy current loss plus hysteresis loss. Hysteresis loss is related to B_r and H_c of Figure 5.2. The area inside a hysteresis loop is the energy lost in each cycle.

Very often the manufacturer of steel laminations furnishes a core loss curve. For a given frequency (e.g., 50 or 60 Hz) the core loss in watts per unit weight or volume is graphed versus the peak B in laminations carrying uniform B. The peak B is the maximum B over a sinusoidal cycle of applied H. Often core loss is roughly proportional to the square of the peak B.

The core loss curve can be used in conjunction with a linear or nonlinear magnetostatic finite element analysis to determine core loss in a device. For example, if B peaks with time at the same instant throughout a device, the finite element program MSC/MAGNETIC calculates total core loss and its distribution by a static analysis of that instant. The program first calculates B in every steel finite element and then uses the core loss curve to determine the associated core loss in every element and the total core loss. If B peaks at different times depending on location within a device, then more instants must be analyzed.

Energies

One verification of any finite element analysis is to show that the summation of all the energies is zero. The energies are those of the appropriate energy functional discussed on page 10.

For example, for electrostatic problems from Eq. (5.4):

$$\int_v \frac{1}{2}\, \epsilon\, E^2\, dv - \int_v \left[\frac{\phi\rho}{2}\right] dv = 0 \qquad (5.50)$$

where the left term is the stored electric energy and the right term is the energy input via charges ρ. The right term does not include energy input by constrained voltage differences.

For linear magnetostatic problems from Eq. (5.11).

$$W_M - W_I = 0 \tag{5.51}$$

where W_M is the energy stored in the magnetic fields. For linear problems,

$$W_M = \int_v \left(\frac{B^2}{2\mu} \right) dv \tag{5.52}$$

and W_I is electrical energy input via currents, which for linear problems is

$$W_I = \int_v \left(\overline{J} \cdot \frac{\overline{A}}{2} \right) dv \tag{5.53}$$

Both W_M and W_I are output by the program MSC/MAGNETIC and MSC/MAGNUM, enabling the user to verify that the solution gives zero total energy.

For nonlinear magnetostatic problems Eq. (5.51) still applies. However, the energy expressions are now given by Eq. (5.12) as

$$W_M = \int_v \left[\int_o^B \overline{H} \cdot d\overline{B} \right] dv \tag{5.54}$$

$$W_I = \int_v \left[\int_o^A \overline{J} \cdot d\overline{A} \right] dv \tag{5.55}$$

Figure 5.4 shows W_M and W_I for a nonlinear problem, as well as the "complementary" or coenergy:

$$W_C = \int_v \left[\int \overline{B} \cdot d\overline{H} \right] dv \tag{5.56}$$

MSC/MAGNETIC and MSC/MAGNUM output W_M and W_C for nonlinear problems by summing over all finite elements. However, it is not possible in one analysis to directly calculate W_I of Eq. (5.55). Instead, W_I of Eq. (5.53) is calculated. Figure 5.4 shows that

$$\int_v \overline{J} \cdot \overline{A} \, dv = W_M + W_c \tag{5.57}$$

Thus, adding W_M and W_c should give twice the energy of Eq. (5.53).

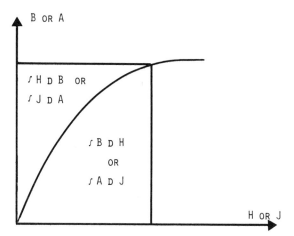

Figure 5.4 Nonlinear *B–H* curve and its energies.

Force

Many electromagnetic devices are designed to produce an electromagnetic force. Calculations of this force aids the design and checks against any available force measurements. In addition to desired forces, undesirable forces may be present and need to be calculated.

A useful general expression for force F is

$$F_x = \frac{\partial W}{\partial x} \tag{5.58}$$

where x is any direction (y, z, or any other direction vector may be substituted), and W is the work or energy producing the force. Equation (5.58) is derived from the definition of work as force times distance, and is called the virtual work expression for force.

An approximation of Eq. (5.58) is

$$F_x(x) = \frac{W(x + \Delta x) - W(x)}{\Delta x} \tag{5.59}$$

where the object on which the force is desired is displaced a small amount Δx in the direction of the unknown force. In finite element analysis the displacement is made by either moving grid points or changing element materials. W is calculated for both positions of the object, and the force is calculated using Eq. (5.59).

Electrostatic forces may be calculated using Eq. (5.59) where W is the stored electric energy of Eq. (5.37) and Eq. (5.50). The energy expressions given are for linear constant permittivity dielectrics but may be altered for nonlinear dielectrics.

An expression for magnetic forces is derived with the aid of Figure 5.5. Two $B–H$ curves are shown for the entire device, including air gaps. Curve 1 is for the first position and curve 2 is for the second position of Eq. (5.59). For both finite element analyses the excitation is kept the same, giving $H = H_A$ in Figure 5.5. The area between the two curves is the energy difference, much like the hysteresis loss as discussed on page 135. Here the energy difference goes into mechanical work.

From the definition of coenergy of Eq. (5.54) and Figure 5.4, the energy difference of Figure 5.5 is the difference in coenergies. Thus, Eq. (5.59) becomes for magnetic forces

$$F_x(x) = \frac{W_C(x + \Delta x) - W_C(x)}{\Delta x} \tag{5.60}$$

where the coenergy W_C is evaluated at two positions with constant excitation. Constant excitation means that applied currents, induced currents, and permanent magnetization maintain their values during the position change.

Occasionally the designer needs to determine the distribution of the above forces. Maxwell's stress tensor [7] may be used to obtain the distribution of

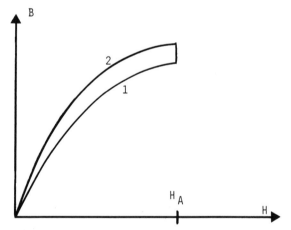

Figure 5.5 Nonlinear $B–H$ relations of a typical magnetic device at two positions (airgaps) and constant current, showing the change in coenergy.

normal and tangential forces on any part of a finite element model. For example, the normal force on a steel-air interface is

$$\frac{F}{S} = \frac{B_N^2}{2\mu} - \frac{B_N^2}{2\mu_o} \tag{5.61}$$

where the steel has constant permeability μ, B_N is normal B, and S is the surface area. Tangential force distributions involve tangential components of B, which may not be calculated accurately at steel-air interfaces unless an extremely fine finite element mesh is used. Experience shows that the method of virtual work is the more accurate method to calculate forces with finite elements.

A special component of magnetic force that is more easily calculated than the above general force is called Ampere's force. This force obeys Ampere's force law:

$$\bar{F} = \int_v \bar{J} \times \bar{B} \, dv \tag{5.62}$$

which becomes, if $\bar{J}$ flows along length ℓ over a surface $\bar{S}$ in which $\bar{B}$ lies,

$$F = B\ell \int \bar{J} \cdot d\bar{S} \tag{5.63}$$

$$F = B\ell I \tag{5.64}$$

This force is calculated by MSC/MAGNETIC for all current-carrying regions. Devices with this force include voice coil actuators for loudspeakers and other applications.

Torque

Torque calculation is important for finite element analysis of electric motors, generators, actuators, and other rotary devices. Torque calculation is very similar to force calculation.

The method of virtual work of Eq. (5.60) becomes for torque T at angle θ

$$T(\theta) = \frac{W_C(\theta + \Delta\theta) - W_C(\theta)}{\Delta\theta} \tag{5.65}$$

where the coenergy W_C must again be calculated at two positions with constant excitation. The angle θ in Eq. (5.65) must be in radians. Typical $\Delta\theta$ values in degrees range from 0.1 to 1.0.

EXAMPLES OF ELECTRIC ANALYSIS

Four practical examples of electric field analysis are discussed below. The examples are for the three problem types discussed on pages 113–115: electrostatic, steady current flow, and electrodynamic. In all four examples the free charge ρ of Eq. (5.1) is zero. All calculations are made using the codes MSC/MAGNETIC or MSC/MAGNUM.

Electrostatic Axisymmetric Insulator

Figure 5.6a shows a porcelain insulator placed between a high-voltage vertical rod and a ground plane. The geometry is axisymmetric about the center line

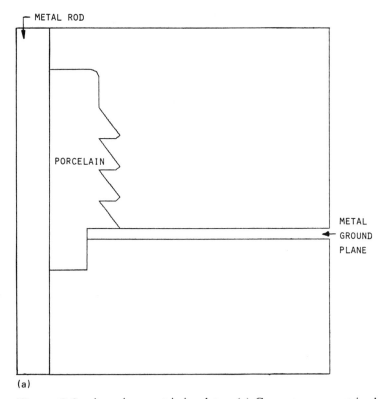

(a)

Figure 5.6 An axisymmetric insulator. (a) Geometry symmetric about left vertical axis. (b) Axisymmetric finite element model. (c) Calculated voltage contours.

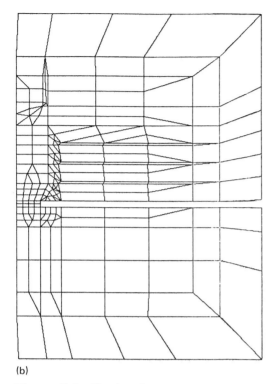

(b)

Figure 5.6 Continued

of the rod. The rod and ground plane are metal. The ground plane is at zero volts and the rod is at 10 kV d.c. or a.c.

Assuming that the porcelain and air have zero electrical conductivity, the calculation of the electric field $\overline{E}$ is an electrostatic problem. Knowledge of $\overline{E}$ is required to determine whether arcing or corona occurs and to determine the capacitance between the rod and ground.

Figure 5.6b shows an axisymmetric finite element model developed for Figure 5.6a. Small finite elements are used in the region directly between the rod and the ground plane, where $\overline{E}$ is expected to be highest and to vary most rapidly with position. The inside of the rod and plane need not be modeled as long as the outer grid points of these electrodes are constrained to 10 kV and 0, respectively. To account for $\overline{E}$ going to infinity the infinite or open boundary elements described on page 127 are used.

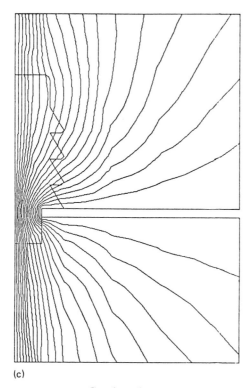

(c)

Figure 5.6 Continued

Figure 5.6c shows the calculated contours of constant voltage. As expected, the contours are closest together directly between the rod and the ground plane, indicating high E at that location. The capacitance can be calculated using Eq. (5.39).

Current Flow in a Fuse

Figure 5.7a shows a fuse made entirely of one metallic material. Electric current flows from the left to the right in the XY plane of Figure 5.7a. The fuse has a known depth in the Z direction. Knowledge of current density distribution $\bar{J}$ and fuse resistance R is needed by the designer.

Figure 5.7b shows a two-dimensional planar finite element model developed for the fuse. Quadrilateral and triangular elements are used. The eight grid points on the left edge are constrained to 1 V and the eight grid points

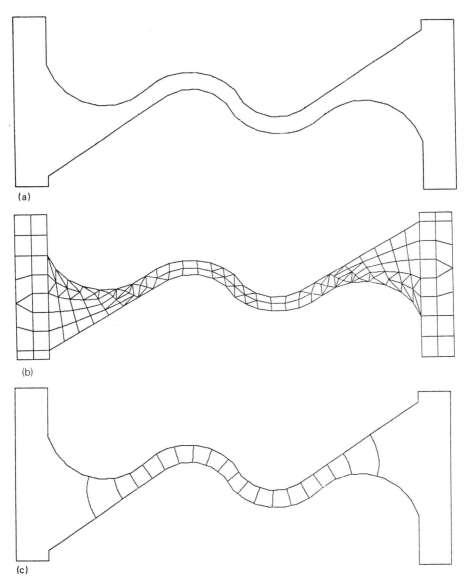

Figure 5.7 A fuse. (a) Cross-section in the plane of current flow. (b) Finite element model. (c) Calculated voltage contours.

on the right edge are constrained to zero. Because no current can leave the other outer boundaries, no constraints are applied to any other grid points. The natural boundary condition of perpendicular voltage contours means from Eq. (5.3) that $\bar{E}$ and $\bar{J}$ do not cross unconstrained boundaries.

Figure 5.7c shows the calculated contours of constant voltage. The related resistance R can be calculated from Eq. (5.43). The calculated power density J^2/σ of Eq. (5.40) can be input to a thermal finite element analysis of Chapter 4 to calculate the temperature distribution. Thus finite element analysis helps the engineer to design the fuse to melt at the desired current.

Electrodynamic Fields in Lossy 3D Capacitor

Figure 5.8a shows a capacitor with two parallel square metal plates. Between the plates are two lossy dielectric materials of equal volume. One has $\epsilon_r = 1$ and $\sigma = 0.333E5$. The other has $\epsilon_r = 1$ and $\sigma = 0.6667E5$. One metal plate is assumed held at 0 V, and the other plate is at a peak a.c. voltage of 1 V of frequency 60 kHz. The voltage distribution in the capacitor is desired.

Figure 5.8b shows a three-dimensional finite element model developed for the capacitor. Because the capacitor geometry is invariant with z, a two-dimensional model would have sufficed. In Figure 5.8b there are eight hexahedral finite elements, four for each of the two lossy dielectrics. The metal

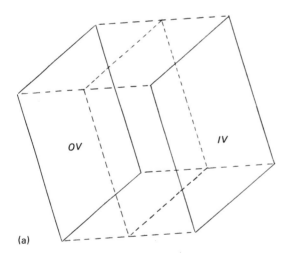

(a)

Figure 5.8 A three-dimensional lossy capacitor. (a) Geometry. (b) Three-dimensional finite element model made of hexahedrons. (c) Calculated voltage contours.

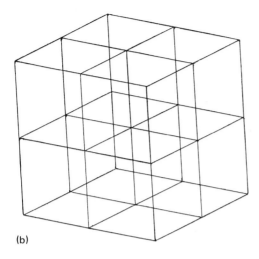

(b)

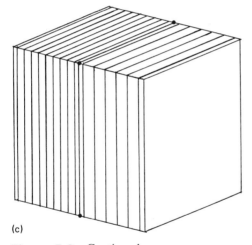

(c)

Figure 5.8 Continued

plates need not be modeled as long their grid points are constrained to the known voltages.

Figure 5.8c shows the calculated contours of constant voltage at time zero (in phase with the applied voltage). The calculated voltage along the grid points between the two dielectrics is $0.3847 + j\, 0.0770$ V. This voltage can be verified by a circuit model consisting of two pure series capacitances in parallel with two different resistances. The circuit approach, however, cannot be used for more complicated geometries, where finite element analysis can still be performed.

Transmission Line

Figure 5.9a shows the cross-section of a two-wire transmission line. The engineer needs to know the capacitance between the two wires, which because of the wire shapes cannot be calculated analytically.

Figure 5.9b shows a two-dimensional finite element model developed for Figure 5.9a. Owing to symmetry only one-half of the line need be modeled, using natural boundary conditions along the boundary of the two halves. The model will also be used for magnetic field calculations on page 168, in which case the finite elements within the wires are required. For the electrostatic calculations of this section all of the left wire grid points are constrained to 1 V and all of the right wire grid points are constrained to zero; thus the inner grid points of the wires may be removed in calculating $\overline{E}$. The wires are separated by air which extends to infinity. Hence infinite elements discussed on page 127 are used along the three outer boundaries.

Figures 5.9c shows the calculated voltage contours. The capacitance C is determined from the stored energy using Eq. (5.39). $\overline{E}$ and C are independent of frequency except perhaps at frequencies above 1 GHz, where the dielectric constants of some materials may vary with frequency.

EXAMPLES OF MAGNETIC ANALYSIS

Twelve practical examples of magnetic field analysis are discussed below. The examples are for the three problem types discussed on pages 115–117: linear magnetostatic, nonlinear magnetostatic, and magnetodynamic. All calculations are made using the finite element program MSC/MAGNETIC or the three-dimensional finite element program MSC/MAGNUM.

d.c. Axisymmetric Nonlinear Solenoid

Figure 5.10a shows a solenoid that is energized by a d.c. current in a stationary stator coil. The coil is wound around an axis of symmetry. The stator also

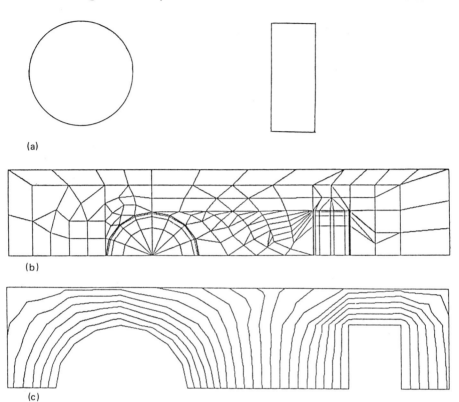

Figure 5.9 Cross-section of a transmission line with current flow normal to the page. (a) Geometry. (b) Finite element model. (c) Calculated voltage contours of electrostatic field.

includes a steel can with a hollow center pole. A hollow movable steel plunger (armature) is attracted to the stator pole by an unknown magnetic force. The designer needs to calculate the force as a function of axial plunger position. To optimize the size, cost, and weight, the designer also needs to calculate the saturable nonlinear $\overline{B}$ throughout the solenoid.

Figure 5.10b shows an axisymmetric finite element model developed for the solenoid. The smallest finite elements are in the small side air gap between the plunger and the bottom of the steel stator. A is constrained to zero on all four boundaries of the model, thereby confining the flux. The steel is described with a nonlinear table of B versus H.

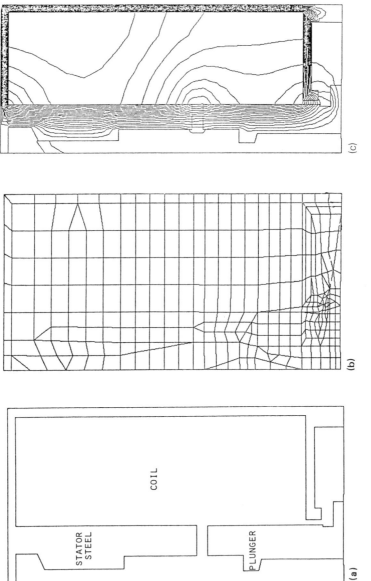

Figure 5.10 d.c. axisymmetric nonlinear solenoid. (a) Geometry symmetric about left vertical boundary. (b) Axisymmetric finite element model. (c) Calculated magnetic flux line pattern.

Figure 5.10c is the calculated magnetostatic flux line pattern for the solenoid when the coil is given a current. A large number of flux lines flow in the coil region, representing leakage and fringing fluxes that are essentially incalculable by any other analysis method. A color display of $\overline{B}$ that is not shown here is also obtained, indicating to the designer where steel should be added or removed to obtain $\overline{B}$ just below saturation for optimum design.

The axial magnetic force on the plunger is calculated by moving the grid points of the plunger a small amount in the axial direction. The change in coenergy determines the force by use of Eq. (5.60). These finite element methods have been shown to greatly aid the design of industrial solenoids and actuators [8].

a.c. Axisymmetric Solenoid

Figure 5.11a shows a solenoid that is energized by a 60-Hz a.c. current in its stator coil. The geometry is axisymmetric and is similar to that of the d.c. solenoid in the preceding section. A new component is the shading ring, which is a copper ring in the middle of the stator pole. The shading ring has an unknown current induced in it that sets up a magnetic field that has a different phase angle than that of the stator coil current. Thus, the magnetic field and force should not go to zero when the applied a.c. current passes through zero 120 times per second.

Figure 5.11b shows an axisymmetric finite element model of the a.c. solenoid. The material properties needed for the magnetodynamic analysis of fields and induced currents are permeabilities and conductivities. Here the permeability of the steel is approximately 2000 times that of air, whereas copper has the same permeability as air. Here the conductivity of the steel is 5.E6 siemens(S)/m and that of copper is 5.8E7 S/m.

The finite elements of the copper shading ring are given the conductivity of copper so that the induced or eddy current in the ring is calculated. However, the finite elements of the coil are given a conductivity of zero. In actuality the coil consists of hundreds of turns of insulated copper wire. If the insulation were removed, there would be significant eddy currents and skin effects in the coil, causing a substantial increase in coil resistance and losses, as will be described on page 168. Modeling the insulation would require thousands of finite elements in the coil region [9]. Instead the model of Figure 5.11b can be used if the coil region is given zero conductivity.

The size of the finite elements in conducting regions should be related to the skin depth in the conducting material. An approximate formula for skin depth δ is

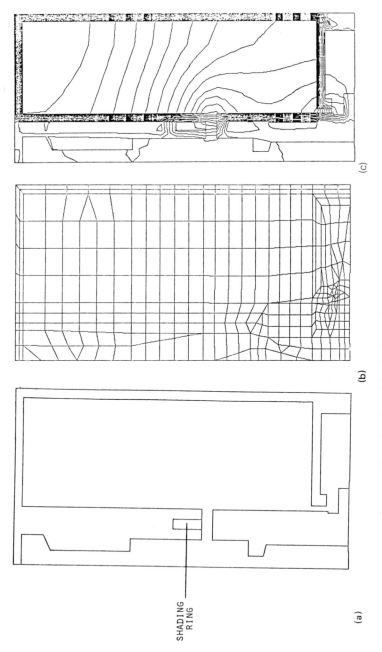

Figure 5.11 a.c. axisymmetric solenoid with shading ring. (a) Geometry axisymmetric about left vertical boundary. (b) Axisymmetric finite element model. (c) Magnetic flux line plot at instant when main coil current peaks.

$$\delta = \frac{1}{\sqrt{\pi f \mu \sigma}} \tag{5.66}$$

where frequency $f = 60$, giving $\delta = 6.50E-4m$ in the steel and $\delta = 8.53E-3m$ in the copper ring. The skin of these materials is modeled with finite elements with depth less than δ in order to accurately calculate the decay of B and induced current over a skin depth.

Figure 5.11c shows the calculated magnetodynamic flux pattern at the instant when the coil current is peaking. The flux lines are concentrated along the outer surface of the plunger and stator pole owing to skin effects. The highest calculated B at any time is below 1.8 T, and thus the assumed steel permeability of 2000 is valid. Because B is very low inside the plunger, the steel there can be removed to give lighter weight and faster plunger motion.

The magnetic force on the plunger is calculated by moving its grid points a small amount. The change in the average stored energy divided by the displacement gives the force averaged over one cycle.

d.c. Brush Motor with Permanent Magnets

Figure 5.12a shows a typical permanent magnet d.c. motor with brushes. The brushes (not shown) are required to feed d.c. current to the coils in the rotor, which is also called the armature. (generator in older books)

One advantage of finite element analysis over other methods of analyzing DC motors is the inherent ability of the finite element method to accurately calculate armature reaction effects. High armature currents have significant effects on flux distribution and torque. Also, armature reaction in permanent magnet machines may permanently demagnetize the magnets. The finite element method can predict the current at which demagnetization occurs. Another inherent advantage of finite element analysis is its ability to calculate torque variation with position, called cogging torque [10].

Figure 5.12b shows the finite element model developed for the motor. Periodic boundary conditions (see p. 125) are applied along the radial boundaries one pole pitch apart.

Figure 5.12c is a flux plot calculated by magnetostatic finite element analysis for a given current in the motor. Armature reaction is visible because the flux density in the left half of the permanent magnet is higher (indicated by the close spacing of the flux lines) than in the right half. The torque can be calculated for any rotor position by rotating the rotor slightly and dividing the change in coenergy by the change in angle, as described on page 139. Both average torque and cogging torque can be obtained as functions of rotor current. Finite element analysis has been used to investigate alternative designs

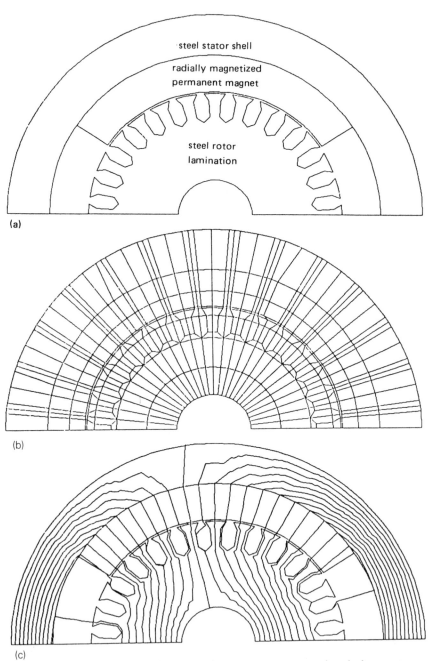

Figure 5.12 One pole of a two-pole permanent magnet brush d.c. motor. (a) Geometry in plane of rotor laminations. (b) Finite element model. (c) Calculated flux line plot.

of this type of motor and has helped achieve a 10-fold reduction in undesirable cogging torque of a particular motor.

Also of great interest is the speed-versus-torque curve of the motor. The speed corresponding to the above-average torque and current can be calculated using the familiar relation for d.c. machines:

$$T * W = Eb * I \qquad \text{applicable to auto generators} \qquad (5.67)$$

where T is torque averaged over 360° of rotation, W is rotational speed in radians per second, Eb is back *EMF* in volts, and I is the armature current. Kirchoff's voltage law gives

$$Es = Eb + I * R \qquad (5.68)$$

where Es is the d.c. source voltage and R is the armature resistance. Substituting Eq. (5.68) in Eq. (5.67) gives

speed rpm

$$W = (Es - I * R) * \left(\frac{I}{T}\right) \qquad (5.69)$$

Hence Eq. (5.69) can be used with the finite element results for torque per ampere to obtain the speed-versus-torque curve of the d.c. motor. The only inaccuracy in the above derivation is due to the neglect of core loss and stray loss in Eq. (5.67). These losses may be significant at extremely high motor speeds.

d.c. Brush Motor with Wound Field

Figure 5.13a shows a typical d.c. motor with brushes and field windings. No permanent magnets are used. The field winding may be in series, in shunt, or compound with the armature (rotor) winding.

Figure 5.13b shows the finite element model developed for the motor of Figure 5.13a. Periodic boundary conditions (see p. 125) are applied to the radial boundaries one pole pitch apart.

The finite element model was submitted to magnetostatic finite element analysis, which produced the flux line plot shown in Figure 5.13c for given field and armature currents. The flux plot is not symmetric about the pole axis owing to armature reaction.

The torque of Figure 5.13c can be calculated from the coenergy in the manner described in the preceding section. The method given there for calculating speed can also be used here for shunt wound motors. Series wound and compound wound motors require changes in Eq. (5.68) and Eq. (5.69) to obtain their speed-versus-torque curves.

See above

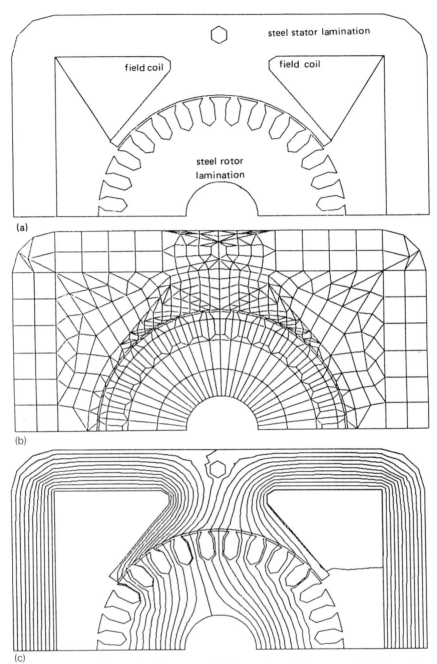

(a)

(b)

(c)

Figure 5.13 One pole of a two-pole d.c. motor with a wound field, or of a two-pole universal motor. (a) Geometry in plane of rotor laminations. (b) Finite element model. (c) Calculated flux line plot.

Universal Motor   *dc motor that work w/Ac* —ᴍ—ᴍ—

The geometry of universal motors can be identical to that of wound field DC motors, such as the one shown in Figure 5.13a. The field and armature windings must be connected in series in order for the motor to produce useful torque when energized by a.c., in which case it is called a universal motor.

Finite element models and flux plots for universal motors are of the type shown previously in Figures 5.13b and c. However, the current and torque now vary with time over the period D of the applied a.c. frequency. The time average torque Tav is calculated from the instantaneous torque $T(i)$ using

$$Tav = \frac{1}{D} \int_o^D T(i)\, dt \tag{5.70}$$

The current i can be any known periodic waveform. Usually it is approximately sinusoidal, but if the universal motor is controlled by solid-state electronic devices, the current may have a square or chopped waveform. For sinusoidal current the torque calculated by the program AOS/MAGNETIC has been shown to agree well with measurement on a highly saturated universal appliance motor [10].

The core losses of both the stator and the rotor can also be calculated from the magnetostatic fields obtained by the finite element method. Calculated losses at various speeds have been shown to agree reasonably well with approximately measurements on a typical universal motor [10].

Brushless Synchronous Motor   *auto alternar*

Figure 5.14a shows a typical synchronous motor with a rotor made of steel and permanent magnets. Similar motors have been analyzed elsewhere [10] and are called either "brushless d.c." or "brushless synchronous" motors. The stator can be excited either with a sinusoidal a.c. current or with a chopped "d.c." or pulse-width-modulated current.

The permanent magnet rotor requires no current, thereby conserving the energy lost in wound coils. Permanent magnet materials have improved greatly over the last few years and are therefore being used more and more. The finite element method can accurately analyze permanent magnets of any shape and material. There is no need to calculate reluctance factors, load lines, or leakage factors if a finite element computer program is used. And there is no need to assume that the magnet is operating at any one point or B value. Instead the magnet's $B-H$ curve is input to the finite element program and the program calculates B and how it varies throughout the magnet and the

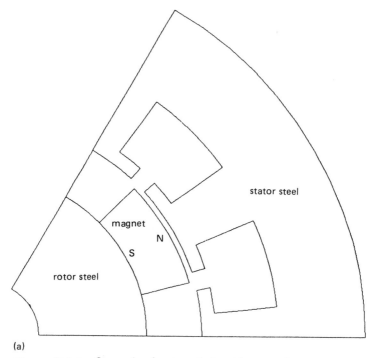

(a)

Figure 5.14   One pole of a six-pole brushless synchronous permanent magnet motor. (a) Geometry in plane of stator laminations. (b) Finite element model. (c) Calculated flux line plot.

entire motor. Finite elements can also aid the design of magnetizing fixtures used in permanent magnet motor manufacture.

Figure 5.14b shows the two-dimensional finite element model of Figure 5.14a. Periodic boundary conditions, as described on page 125, are applied along the radial boundaries one pole pitch apart.

Figure 5.14c shows the saturable magnetic flux distribution calculated by magnetostatic finite element analysis. Note that owing to armature reaction (stator current) the flux pattern is not symmetric about the magnet axis. The flux density also varies noticeably within the permanent magnet.

The torque may be calculated from the change in total coenergy divided by change in rotor angle. The torque at various rotor positions can be determined. Torque variation over a stator slot pitch is called cogging torque. The average torque (over all rotor positions) is the torque that does useful work

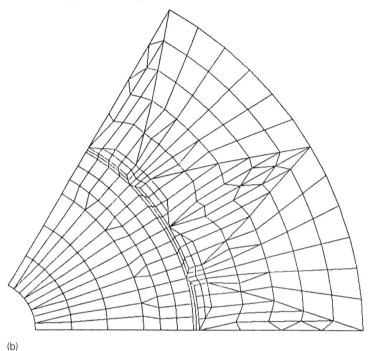

(b)

Figure 5.14 Continued

and is commonly measured on a dynamometer. Agreement within 5% between measurement and finite element calculation has been found for permanent-magnet motors [11].

Stepper Motor

A stepper motor (also called a step motor or stepping motor) is a synchronous machine that is excited by current pulses of variable frequency. It can be "stepped" pulse by pulse to any desired incremental position [12].

One common type of stepper motor is the variable reluctance (VR) stepper. Figure 5.15a shows a typical VR motor, which has essentially planar two-dimensional magnetic fields. The unexcited rotor tends to rotate to positions dependent on which stator coils are excited. When the rotor speed is high owing to rapid continual pulse excitation, the motor is commonly called a switched reluctance motor.

Figure 5.15b shows the saturable magnetostatic flux pattern calculated by MSC/MAGNETIC. The torque calculated by the change in coenergy is within

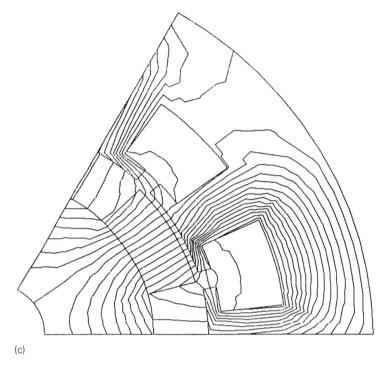

(c)

Figure 5.14   Continued

a few percent of the measured torque. For accurate torque calculation it is found that air gap finite elements need to have radial edges at the end of each stator or rotor tooth. Other types of stepper motors have also been analyzed by finite elements. Analysis of a hybrid stepper motor containing both permanent magnets and coils of wire has been described elsewhere [12].

Recording Head

Magnetic recording heads are used in magnetic tape recorders and magnetic disk drives to write and read signals on magnetic media. Figure 5.16a shows a typical recording head positioned above a thin medium such as a disk or tape. Figure 5.16b shows a finite element model developed for the region.

Assuming that the medium is initially unmagnetized, calculation of the magnetic fields is fairly straightforward. Given the magnetization curves of quadrant 1 of Figure 5.2 for both the head material and the medium material, the nonlinear saturable magnetostatic field is calculted during magnetization as shown in Figure 5.16c. This field is useful to the designer in several ways.

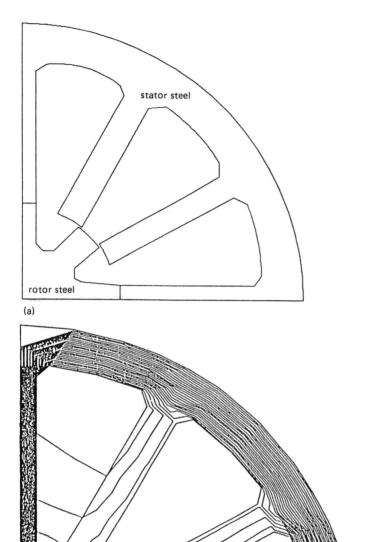

(a)

(b)

Figure 5.15 One-quarter of a variable reluctance (or switched reluctance) stepper motor. (a) Geometry in plane normal to the shaft. (b) Calculated flux line plot.

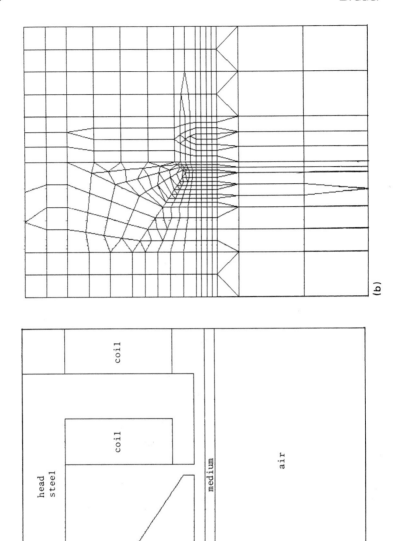

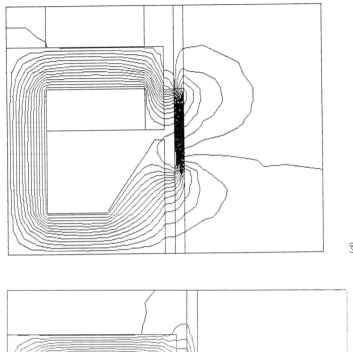

(c)

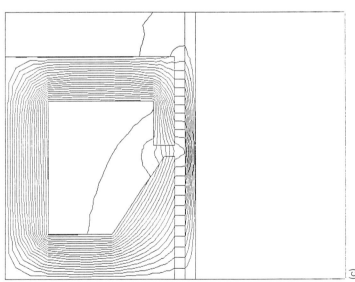

(d)

Figure 5.16 Magnetic recording head and recording medium. (a) Geometry in plane parallel to the motion of the medium. (b) Finite element model. (c) Calculated flux line plot due to current in head coil. (d) Calculated flux line plot due to magnetized medium.

Knowledge of saturated regions of the head (such as the left pole tip) helps determine head shape for optimum material usage for a given coil current. Knowledge of the field in the medium helps the designer obtain highest density of data storage.

After the head coil current is removed, the resulting nonlinear magnetostatic field due to the magnetized medium can be calculated using one of several algorithms [14]. The magnetization of each finite element in the medium can be different in magnitude and direction. Figure 5.16d shows a typical flux pattern due to the magnetized medium. B is much lower than in Figure 5.16c, meaning that the effective head gap may be smaller for readback than for recording, which is usually desirable.

Transformer

Transformers of all types can be analyzed by magnetic finite elements. The types include high-power and low-power single- and multiphase, high-frequency switchers, and ferroresonant transformers.

Figure 5.17a shows a typical low-power single-phase transformer. As in all transformers, there are two windings called the primary and the secondary. The primary is connected to a voltage source and always carries some current, and the secondary current depends on the load connected to the secondary terminals. Figure 5.17b shows a finite element model developed for the transformer.

Figure 5.17c shows the magnetostatic flux pattern for current in the primary only. The secondary has zero current and therefore has an open circuit load. Under these conditions basic transformer theory [15] shows that the inductance seen by the primary is the saturable magnetizing inductance Lmag as shown in the transformer equivalent circuit of Figure 5.18, where Lmag is much greater than Lleak. The corresponding magnetizing reactance is

$$Xmag = \omega \, Lmag \tag{5.71}$$

Because of saturation, Lmag can be a function of primary current and is determined from Figure 5.17c for various primary currents.

To determine no-load primary current as a function of time for a given voltage waveform $V(t)$, Faraday's law is applied to Figure 5.17c:

$$V = -\frac{d\lambda}{dt} \tag{5.72}$$

where λ is flux linkage discussed on page 132. Assuming the primary resistive drop IR_1 of Figure 5.18 is negligible gives

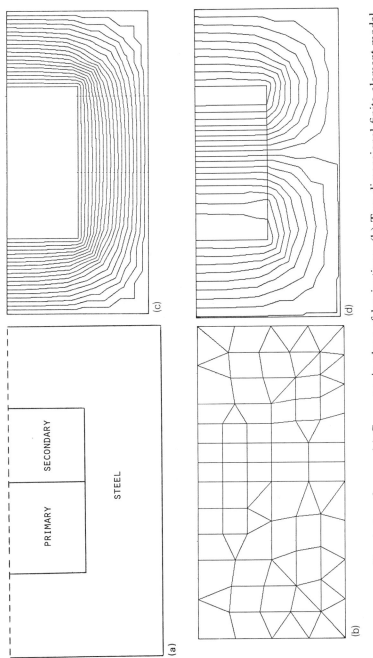

Figure 5.17 Simple transformer. (a) Geometry in plane of laminations. (b) Two-dimensional finite element model. (c) Calculated flux line plot due to primary current only. (d) Calculated flux line plot due to equal and opposite primary and secondary currents.

$$\lambda(t) = -\int V(t)\, dt \tag{5.73}$$

Then $\lambda(I)$ of the finite element output determines the current waveform $I(t)$ [16].

The other important inductance of any transformer is the leakage inductance Lleak. The transformer equivalent circuit of Figure 5.18 shows that Lleak is obtained by short-circuiting the secondary, in which case the total secondary ampere-turns are approximately equal and opposite to those of the primary. Hence secondary amp-turns opposite the primary amp-turns are input to the finite element model, giving the magnetostatic flux pattern shown in Figure 5.17d. The inductance associated with this leakage flux pattern is the total leakage inductance Lleak.

For secondary loads other than short or open circuits, the transformer equivalent circuit of Figure 5.18 made up of the above magnetizing and leakage inductances can be used. Thus the transformer currents and voltages can be determined under any load.

Losses in the transformer can also be determined from finite element analysis. Steel core losses are output by using the core loss curve for the steel. Eddy current losses in the windings can be estimated for wire sizes smaller than a skin depth by using simple formulae based on B in the windings. Calculation of eddy current losses for wire diameters bigger than one skin depth requires modeling of individual wires, as shown on page 168.

Double-Cage Induction Motor

As explained in books on motors [15], an induction motor is basically a transformer with a moving secondary called a squirrel cage rotor. The primary is stationary and is called the stator.

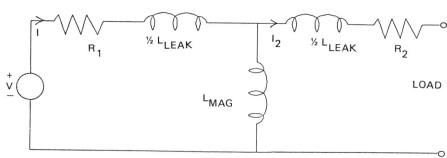

Figure 5.18   Equivalent electric circuit of a transformer.

One way of analyzing induction motors is called the equivalent circuit method. The equivalent circuit is very similar to that in Figure 5.18 and again contains two basic inductances and reactances: magnetizing and leakage. Thus the techniques of the preceding transformer section can be used to calculate these saturable reactances accurately, as shown in detail elsewhere [13]. An exception is the double-cage induction motor, in which the rotor current distribution is unknown owing to skin effects.

To calculate the rotor current distribution, the magnetodynamic finite element method described on page 117 can be used. The conductivity of the rotor bars and the frequency seen by the rotor are input to the program, from which the induced rotor currents and their effects on B, the impedance, and the torque can be calculated.

Figure 5.19a shows one-quarter of a four-pole polyphase induction motor. There are 24 stator slots and 32 double-cage rotor bars in the entire motor. Figure 5.19b shows the finite element model.

Figure 5.19c shows the flux line plot calculated for the motor at start, for which the frequency seen by the rotor is the line frequency of 60 Hz. The flux penetrating the rotor is much less than in Figure 5.19d, which is at high speed with low frequency seen by the rotor. The calculated resistive component of impedance thus is much higher at low speed than at high speed, which is desirable for good motor performance. An alternative way of calculating solely the rotor resistance and inductance as a function of frequency by modeling just one rotor bar is similar to the method described on page 168.

Automotive 3D Alternator

Several kinds of motors and alternators have highly three-dimensional flux paths. The flux flows not only in the plane normal to the shaft axis, but also in the axial direction. There are two ways to analyze such devices with finite elements.

One way is to perform one or more planar and/or axisymmetric two-dimensional analyses of the device. Certain approximations are involved, but sometimes results can be obtained fairly quickly and accurately. Examples are a novel axial flux stepper motor analysis with three two-dimensional models and a hybrid stepper motor analysis [12].

The second way of analyzing three-dimensional problems is to use three-dimensional finite elements. This technique avoids the approximations involved in developing and/or combining two-dimensional models and shows the true three-dimensional flow of flux. Here three-dimensional analysis is applied to

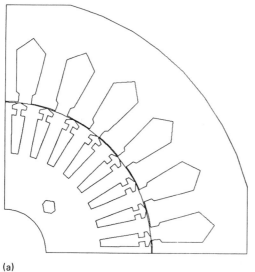

(a)

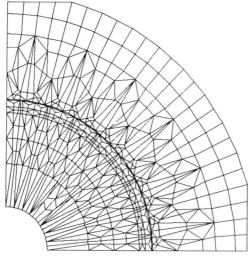

(b)

Figure 5.19   One pole of a four-pole polyphase induction motor with double-cage rotor. (a) Geometry in plane of laminations, normal to shaft. (b) Two-
dimensional finite element model. (c) Calculated flux line plot at start.
(d) Calculated flux line plot at high rotor speed.

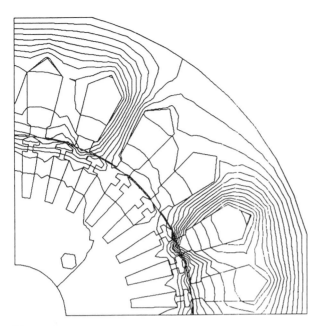

(c)

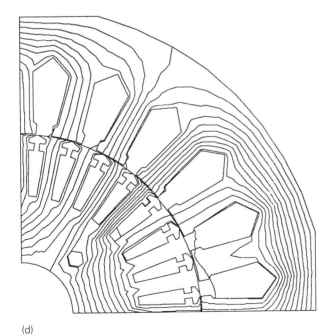

(d)

Figure 5.19 Continued

the Lundell or claw-pole-type alternator that is commonly used in automobiles [17].

Figure 5.20a shows a sketch of one pole pitch of a 12-pole automotive alternator. Figure 5.20b shows portions of a corresponding three-dimensional finite element model. Approximately 1800 finite elements are used in the model, consisting of hexahedrons, pentahedrons, and tetrahedrons.

The three-dimensional model of Figure 5.20b was input to MSC/MAGNUM, a three-dimensional electromagnetic program [4], which calculated the static A and B distributions throughout the machine for the rotor position shown in Figure 5.20a. Plotting the magnitude of A gives the flux line plot of Figure 5.20c for the no-load case of rotor current without stator current. Color displays of B may also be obtained.

In the no-load case shown in Figure 5.20c the flux is confined to the modeled pole pitch and thus the tangential components of A are set to zero along all boundary planes and cylindrical sectors. In cases with stator current "loads" the flux is not necessarily confined to one pole pitch and MPCs are often used along the boundary planes one pole pitch apart.

Calculated fluxes, inductances, and voltages for various stator output currents have been shown to agree well with measurements. MSC/MAGNUM has been extensively used to optimize the design of various sizes of automotive and truck alternators [17].

Transmission Line

The transmission line analyzed for its electric fields on page 146 is here analyzed for its magnetic fields. The finite element model of Figure 5.9b is used with the permeabilities and conductivities of the steel wires and air. The finite elements near the conductor surfaces are less than one skin depth in radial depth at the highest frequency to be analyzed.

Figure 5.21a is the calculated magnetic flux plot at DC. Figure 5.21b is the calculated magnetodynamic flux plot at 1 kHz, showing the concentration of flux lines near the conductor surfaces. The total current in each conductor is the sum of the applied current and the induced current and also is calculated to be concentrated near the surface of each conductor [9]. These skin effects cause the calculated resistance to increase with frequency, and the calculated inductance to decrease with frequency. The total current in one conductor must be equal and opposite to that in the other conductor, according to Kirchoff's law.

Figure 5.22 is the equivalent circuit of the transmission line. The resistance R and inductance L per meter length are obtained as functions of frequency

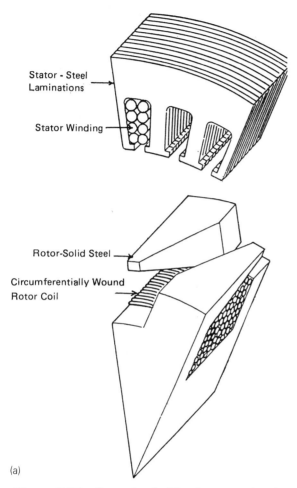

Stator - Steel
Laminations

Stator Winding

Rotor-Solid Steel

Circumferentially Wound
Rotor Coil

(a)

Figure 5.20   One pole of a 12-pole automotive alternator. (a) Three-dimensional sketch. (b) Portions of three-dimensional model. (c) Typical plot of contours of constant magnitude of A.

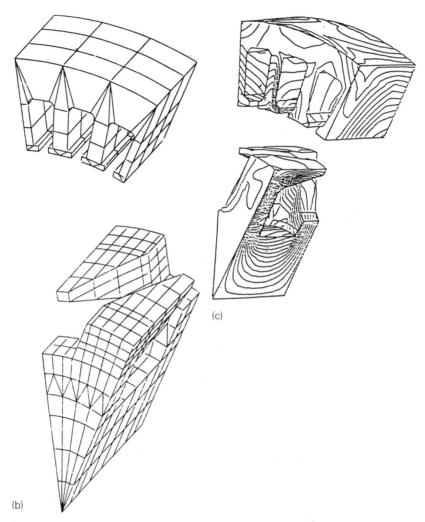

(c)

(b)

Figure 5.20   Continued

from the magnetic fields of Figure 5.21. The capacitance is obtained from the previous electrostatic analysis. The shunt conductance is here assumed zero because of the air between the conductors. If a noninsulating material lies between the two conductors, then the current flow solution described on page 114 can be used to determine the resistance between them for use as the reciprocal of conductance in Figure 5.22. Figure 5.22 is the equivalent

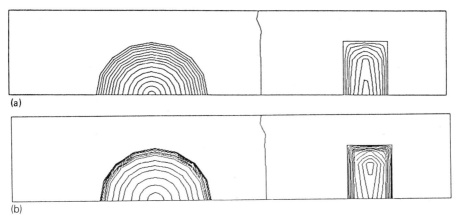

Figure 5.21 Calculated magnetic flux lines in transmission line of Figure 5.9a. (a) For d.c. current. (b) For currents of frequency 1 kHz.

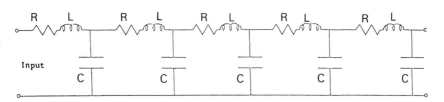

Figure 5.22 Equivalent electric circuit of transmission line.

transverse electromagnetic (TEM) transmission line circuit for the transmission line. It can be used to determine the behavior of any system that contains the line and is useful in studies of electromagnetic compatibility of electrical systems.

EXAMPLES OF ELECTROMAGNETIC ANALYSIS

Microwave Cavity at Resonance

In the preceding section TEM fields were calculated separately as electric fields and magnetic fields. For true electromagnetic problems as described on page 118, the electric and magnetic fields interact and cannot be calculated separately. The example analyzed here of coupled electromagnetic fields is transverse electric (TE) resonance in a microwave cavity such as a microwave oven.

Figure 5.23 shows a finite element model of one-quarter of a cubic box 1 m on each side. The other three-quarters of the box (cavity) are in the remaining three quadrants of *xz* space. The boundaries of the cavity (the walls of the box) are good metallic conductors, giving tangential electric field of zero. Thus tangential A is set to zero along the box walls, which are three surfaces in Figure 5.23. As discussed on page 128, B is thus tangential to those three surfaces, which is known to be true because the induced wall currents prevent B from passing through.

The cavity is excited by a current density of 1 amp/m^2 impressed by a probe in the Y direction in hexahedron number 1 of Figure 5.23. The cavity interior is given a slightly lossy conductivity of 1.E − 4S/m. The frequency of the current was varied and the fields and energies were calculated by three-dimensional electromagnetic finite element analysis.

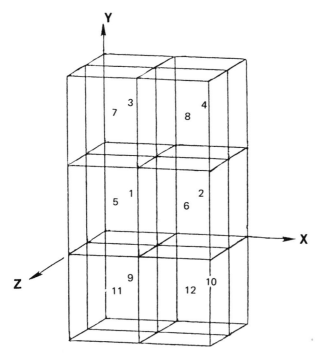

Figure 5.23 Three-dimensional finite element model of one-quarter of a microwave cavity resonator.

The calculated fields and energies were found to peak at a frequency of 223 mHz. This resonant frequency can be compared with the classical equation for the natural frequency of the fundamental TE mode of a rectangular cavity [7]:

$$f = \left(\frac{c}{2}\right) \sqrt{\left(\frac{1}{d}\right)**2 + \left(\frac{1}{a}\right)**2} \tag{5.74}$$

where c = velocity of light and a and d are the cavity dimensions in the X and Z directions, respectively. Here $a = d = 1$, giving $f = 212.13$ mHz. The calculated 223 mHz of the 12 finite element model of Figure 5.23 is in reasonably good agreement. It can be shown that as the number of finite elements is increased, the calculated resonant frequency rapidly approaches 212.13 mHz.

Additional results calculated with the finite element model of Figure 5.23 are graphed in Figure 5.24. It is seen that the calculated electric and magnetic fields agree very closely with the classical theory of sinusoidal TE101 fields in the cavity [7]. The other components of E and B are negligibly small.

Another check of the finite element calculations is made using the classical equation [7]:

$$\frac{E_{y\max}}{B_{x\max}} = i\,2\,f\,d \tag{5.75}$$

Using $d = 1$ and f from Eq. (5.74) gives

$$\frac{E_{y\max}}{B_{x\max}} = i\,424.264\text{E6} \tag{5.76}$$

In comparison, the finite element results of Figure 5.23 gives

$$\frac{E_{y\max}}{B_{x\max}} = \frac{-6104}{i1.44367\text{E}-5} = i\,422.811\text{E6} \tag{5.77}$$

which agrees with Eq. (5.76) within 0.34%.

The advantage of the finite element method is that cavities of any shape, size, and material content can be analyzed, which cannot be accomplished by classical analytical methods.

Microwave Cavity Off-Resonance

The microwave cavity of the preceding section can also be analyzed at frequencies other than resonance. Such analyses are difficult or impossible to accomplish by classical techniques.

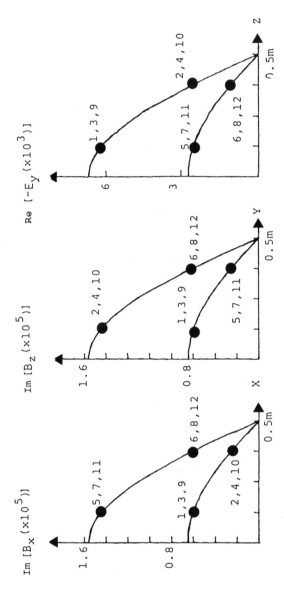

Figure 5.24 Comparison of fields obtained by Figure 5.23 (data points) with classical sinusoidal curves.

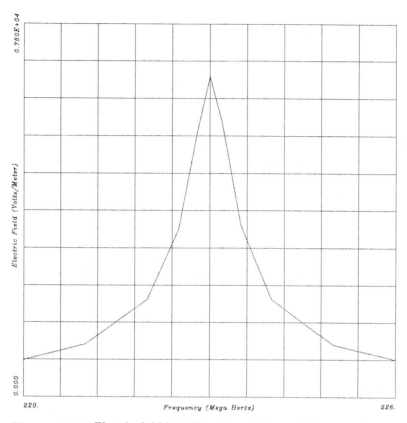

Figure 5.25 Electric field in *Y* direction of Figure 5.23 versus frequency in mHz.

Figure 5.25 is a graph of the peak electric field *Ey* in the cavity versus frequency. The resonant peak at 223 mHz is clearly seen. The width of the peak is determined by the conductivity of the material inside the cavity. For the given conductivity of 1.E − 4 S/m, *E* decreases with frequency as shown.

REFERENCES

1. Brauer, J. R., "Saturated Magnetic Energy Functional for Finite Element Analysis of Electric Machines," IEEE Power Engineering Society Meeting, paper C75-151-6, January 1975.

2. Silvester, P., H. Cabayan, and B. Browne, "Efficient Techniques for Finite Element Analysis of Electric Machines," *IEEE Trans., PAS-92:* 1274–1281, July 1973.

3. Brauer, J. R., "Finite Element Analysis of Electromagnetic Induction in Transformers," IEEE Power Engineering Society Meeting, paper A77-122-5, February 1977.

4. A proprietary product available from MacNeal-Schwendler, 9076 N. Deerbrook Trail, Milwaukee, WI 53223, USA.

5. A proprietary product available from MacNeal-Schwendler, 9076 N. Deerbrook Trail, Milwaukee, WI 53223, USA.

6. Brauer, J. R., "Open Boundary Finite Elements for Axisymmetric Magnetic and Skin Effect Problems," *J. Appl. Phys., 53:* 8366–8368, November 1982.

7. Plonsey, R., and R. E. Collin, *Principles and Applications of Electromagnetic Fields,* McGraw-Hill Book Co., New York, 1981.

8. Hastings, J. K., M. A. Juds, and J. R. Brauer, "Accuracy and Economy of Finite Element Magnetic Analysis," in *Proceedings of National Relay Conference,* April 1985.

9. Brauer, J. R., "Finite Element Calculations of Eddy Currents and Skin Effects," *IEEE Trans., MAG-18:* 504–509, March 1982.

10. Brauer, J. R., "Finite Element Calculation of Synchronous, Universal, and Induction Motor Performance," *MOTORCON Proc.,* March 1982.

11. Brauer, J. R., L. A. Larkin, and V. D. Overbye, "Finite Element Modeling of Permanent Magnet Devices," *J. Appl. Phys. 55:* 2183–2185, March 1984.

12. Brauer, J. R., "Finite Element Analysis of DC Motors and Step Motors," *Proceedings of Incremental Motion Control Symposium,* May 1982.

13. Brauer, J. R., "Finite Element Analysis of Single-phase and Poly-phase Induction Motors," *Conference Record of IEEE Industry Application Society Meeting,* October 1981.

14. Koehler, T. R., "Self-Consistent Vector Calculation of Magnetic Recording Using the Finite Element Method," *J. Appl. Phys. 55:* 2214–2216, March 1984.

15. Englemann, R. H., *Static and Rotating Electromagnetic Devices,* Marcel Dekker, Inc., New York, 1982.

16. Silvester, P., and M. V. K. Chari, "Finite Element Solution of Saturable Magnetic Field Problems," *IEEE Trans., PAS-89:* 1642–1652, September 1970.

17. Zeisler, F. L., and J. R. Brauer, "Automotive Alternator Electromagnetic Calculations Using Three Dimensional Finite Elements," *IEEE Trans., MAG-21:* 2453–2456, November 1985.

6 Fluid Analysis

Nancy J. Lambert

The MacNeal-Schwendler Corporation, Milwaukee, Wisconsin

PROBLEM TYPES

Since a large portion of the earth is in a fluid state, it is of interest to engineers and scientists to study the behavior of fluids. The many applications of fluid mechanics make it one of the most fundamental of all engineering studies. The study of fluid machinery such as pumps, compressors, heat exchangers, and engines is of importance to mechanical engineers. The flow of air over objects is of interest to aeronautical engineers in the design of aircraft, missiles, and rockets. The study of the ocean and atmosphere is of importance to oceanography, meteorology, and hydrology. Fluid mechanics and electromagnetic theory are combined in magnetohydrodynamics. Acoustics is a combination of structural and fluid mechanics.

A fluid is a substance in which the molecules are able to flow past each other without limit and without forming fracture planes. Fluids are usually classified as gases or liquids. Liquids have only slight compressibility and their density varies little with temperature or pressure. On the other hand, for a given mass of gas, the pressure, temperature, and enclosing volume are related by the appropriate equation of state of the gas.

Fluid flow models commonly handled by the finite element method can be classified into several broad categories [1], as shown in Table 6.1. The range of problems encountered in fluid mechanics is very large. One technique

Table 6.1 Finite Element Fluid Flow Models

Model	Reynolds Number	Governing Equation
Ideal, irrotational, incompressible	∞	Euler's momentum equation, conservation of mass
Viscous, incompressible without inertia	0	Stokes' (creeping) flow
Viscous, incompressible with inertia effects	0 to ∞	Navier-Stokes equation
Viscous, compressible with inertia effects	0 to ∞	Navier-Stokes equation with compressibility terms

Table 6.2 Steps in Finite Element Fluid Methods

1. Define the governing equations.
2. Define the boundary and initial conditions.
3. Divide the flow domain into finite elements.
4. Define a spatial variation for the variables.
5. Derive a discrete form of the governing equation.
6. Form the matrices.
7. Solve for the unknowns.

will be presented for solving steady-state incompressible flow in the form of the Navier-Stokes equations. The variables of velocity and pressure will be used. This will form a basis for many of fluid flow applications. The interested reader may pursue other references [2–10].

The essential steps in the application of the finite element method are defined in Table 6.2. These steps will be presented in the subsequent sections.

Inviscid Potential Flow

Fluid motion around objects can be divided into two regions: a thin region close to the object where frictional effects are important, and an outer region where friction is negligible. This section is concerned with flows of negligible friction. The flow outside the boundary layer is frictionless and irrotational and is known as potential flow. By potential flow we mean that the velocity of the fluid is derived from the scalar velocity potential ϕ as

$$\overline{V} = -\nabla\phi \tag{6.1}$$

Incompressible potential flow theory is valid for subsonic flow where the Mach number M is less than about 0.3 [11]. When the fluid is incompressible, the velocity in terms of the potential ϕ may be substituted into the continuity equation $\nabla\cdot\overline{V} = 0$ to yield the condition that ϕ is harmonic satisfying Laplace's equation:

$$\nabla^2 \phi = 0 \tag{6.2}$$

In Cartesian coordinates this is

$$\frac{\partial^2\phi}{\partial x^2} + \frac{\partial^2\phi}{\partial y^2} + \frac{\partial^2\phi}{\partial z^2} = 0 \tag{6.3}$$

Applying the method of weighted residuals such that

$$\int_v \text{weighting function} \times \text{differential equation} = 0 \tag{6.4}$$

we have

$$\int_v N_i \left(\frac{\partial^2\phi}{\partial x^2} + \frac{\partial^2\phi}{\partial y^2} + \frac{\partial^2\phi}{\partial z^2}\right) dx\, dy\, dz = 0 \tag{6.5}$$

This results in the equation [2] for two dimensions as

$$\iint_\Omega \left(\frac{\partial N_i}{\partial x}\frac{\partial\phi}{\partial x} + \frac{\partial N_i}{\partial y}\frac{\partial\phi}{\partial y}\right) d\Omega = 0 \tag{6.6}$$

Introducing trial functions to depict the spatial variation of ϕ we have, for each quadrilateral element with eight grid points [2] shown in Figure 6.1,

$$\iint_\Omega \left(\frac{\partial N_i}{\partial x}\sum_{j=1}^{j=8}\frac{\partial N_i}{\partial x}\phi_j + \frac{\partial N_i}{\partial y}\sum_{j=1}^{j=8}\frac{\partial N_j}{\partial y}\phi_j\right) dx\, dy \tag{6.7}$$

This can be written in matrix form as

$$[A]\{\phi\} = [F] \tag{6.8}$$

in which A is the coefficient matrix and F the flux vector:

$$a_{ij} = \iint_\Omega \left(\frac{\partial N_i}{\partial x}\frac{\partial N_j}{\partial x} + \frac{\partial N_i}{\partial y}\frac{\partial N_j}{\partial y}\right) d\Omega \tag{6.9}$$

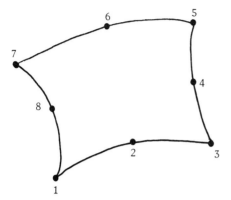

Figure 6.1   Parabolic eight-noded element.

and

$$F_i = \iint_{\angle} Ni \, \frac{\partial \phi_i}{\partial n} \, d\angle \tag{6.10}$$

This gives an 8×8 symmetric matrix, where $A_{ij} = A_{ji}$.

As the flow speed increases, compressibility effects become important. As the Mach number passes unity, shock waves can form on the body. When the Mach number of the free stream is greater than unity ($M > 1.0$), the flow is supersonic, and when the Mach number is less than unity ($M < 1.0$), the flow is subsonic. For M greater than about 6 the flow is hypersonic.

The laws governing flow are basically the same for subsonic through hypersonic flow. The various terms in the fundamental equations become more or less dominant as M changes. The entire character of the flow pattern drastically changes for subsonic through supersonic flow.

Viscous Flow

Most fluids exhibit viscous behavior. This causes energy dissipation which affects the motion of fluids. The governing equations of motion of a compressible viscous fluid are the conservation of mass and conservation of momentum [11]. The equation of continuity assumes the following form:

$$\frac{\partial \rho}{\partial t} + \frac{\partial}{\partial x} (\rho u) + \frac{\partial}{\partial y} (\rho v) + \frac{\partial}{\partial z} (\rho w) = 0 \tag{6.11}$$

where ρ is density, t is time, and flow velocity components u, v, w are in the x, y, and z directions, respectively.

The Navier-Stokes equations are [11]

$$\rho \frac{Du}{Dt} = X - \frac{\partial p}{\partial x} + \frac{\partial}{\partial x}\left[\mu\left(2\frac{\partial u}{\partial x} - \frac{2}{3}\operatorname{div} w\right)\right]$$
$$+ \frac{\partial}{\partial y}\left[\mu\left(\frac{\partial u}{\partial y} + \frac{\partial v}{\partial x}\right)\right] + \frac{\partial}{\partial z}\left[\mu\left(\frac{\partial w}{\partial x} + \frac{\partial u}{\partial z}\right)\right] \tag{6.12}$$

$$\rho \frac{Dy}{Dt} = Y - \frac{\partial p}{\partial y} + \frac{\partial}{\partial y}\left[\mu\left(2\frac{\partial v}{\partial y} - \frac{2}{3}\operatorname{div} w\right)\right]$$
$$+ \frac{\partial}{\partial z}\left[\mu\left(\frac{\partial v}{\partial z} + \frac{\partial w}{\partial y}\right)\right] + \frac{\partial}{\partial x}\left[\mu\left(\frac{\partial u}{\partial y} + \frac{\partial v}{\partial x}\right)\right] \tag{6.13}$$

$$\rho \frac{Dw}{Dt} = Z - \frac{\partial p}{\partial z} + \frac{\partial}{\partial z}\left[\mu\left(2\frac{\partial w}{\partial z} - \frac{2}{3}\operatorname{div} w\right)\right]$$
$$+ \frac{\partial}{\partial x}\left[\mu\left(\frac{\partial w}{\partial x} + \frac{\partial u}{\partial z}\right)\right] + \frac{\partial u}{\partial y}\left[\mu\left(\frac{\partial v}{\partial z} + \frac{\partial w}{\partial y}\right)\right] \tag{6.14}$$

where X, Y, and Z are external body forces in the x, y, and z directions, respectively, and μ is absolute viscosity.

The unknowns in the Navier-Stokes equations and the continuity equation are the velocities u, v, w, and the pressure p.

Let us consider the general steady-state, incompressible Navier-Stokes equation in two dimensions. Writing the Navier-Stokes equations in dimensionless form facilitates generalization for a large range of problems. The following dimensionless variables are used:

$$x^* = \frac{x}{1}, \qquad y^* = \frac{y}{1}, \qquad u^* = \frac{u}{u_o}, \qquad v^* = \frac{v}{u_o} \tag{6.15}$$

and

$$p^* = \frac{p}{\rho u_o^2} \tag{6.16}$$

where 1 is a characteristic length and u_o a datum velocity.

A dimensionless form of the steady-state, incompressible Navier-Stokes equations [2] is

$$u^* \frac{\partial u^*}{\partial x^*} + v^* \frac{\partial u^*}{\partial y^*} = \frac{F_x 1}{\rho u_o^2} - \frac{\partial p^*}{\partial x^*} + \frac{\mu}{\rho u_o\, 1}\left(\frac{\partial^2 u^*}{\partial x^{*2}} + \frac{\partial^2 u^*}{\partial y^{*2}}\right) \tag{6.17}$$

$$u^* \frac{\partial v^*}{\partial x^*} + v^* \frac{\partial v^*}{\partial y^*} = \frac{F_y 1}{\rho u_o^2} - \frac{\partial p^*}{\partial y^*} + \frac{\mu}{\rho u_o 1}\left(\frac{\partial^2 v^*}{\partial x^{*2}} + \frac{\partial^2 y^*}{\partial y^{*2}}\right) \tag{6.18}$$

where F_x and F_y are gravitational body forces per unit volume in the x and y direction, respectively.

The corresponding continuity equation becomes [2]

$$\frac{\partial u^*}{\partial x^*} + \frac{\partial v^*}{\partial y^*} = 0 \tag{6.19}$$

We introduce the following to further simply the equations:

$$R_E = \frac{\rho u_o 1}{\mu} \quad \text{— Reynolds number} \tag{6.20}$$

and

$$F_R = \frac{u_o}{g1} \quad \text{— Froude number} \tag{6.21}$$

where g is gravitational acceleration.

This results in the final form where the asterisk is dropped from the equations and it is assumed that both the velocities and pressures are dimensionless:

$$u\frac{\partial u}{\partial x} + v\frac{\partial u}{\partial y} = \frac{1_{x1}}{F_R^2} - \frac{\partial p}{\partial x} + \frac{1}{R_E}\left(\frac{\partial^2 u}{\partial x^2} + \frac{\partial^2 u}{\partial y^2}\right) \tag{6.22}$$

$$u\frac{\partial v}{\partial x} + v\frac{\partial v}{\partial y} = \frac{1_{x2}}{F_R^2} - \frac{\partial p}{\partial y} + \frac{1}{R_E}\left(\frac{\partial^2 v}{\partial x^2} + \frac{\partial^2 v}{\partial y^2}\right) \tag{6.23}$$

where 1_{x1} and 1_{x2} are direction cosines of the x and y global axes to the direction of the gravitational field.

The continuity equation becomes

$$\frac{\partial u}{\partial x} + \frac{\partial v}{\partial y} = 0 \tag{6.24}$$

Applying the method of weighted residuals over ne elements [2]

$$\sum_1^{ne} \int_{A_e} N_i \left[\sum_1^n N_k u_k \sum_1^n \frac{\partial N_j}{\partial x} u_j + \sum_1^n N_k v_k \sum_1^n \frac{\partial N_j}{\partial y} u_j + \sum_1^m \frac{\partial M_1}{\partial x} p_1 \right.$$

$$\left. - \frac{1_{x1}}{F_R^2} - \frac{1}{R_E}\left(\sum_1^n \frac{\partial^2 N_j}{\partial y^2} u_j + \sum_1^n \frac{\partial^2 N_j}{\partial y^2} u_j\right) \right] dA^e = 0 \tag{6.25}$$

This results in the momentum equation [2] in the x direction as

$$\sum_1^{ne} \int_{A^e} \left[N_i N_k u_k \frac{\partial N_j}{\partial x} u_j + N_i N_k v_k \frac{\partial N_j}{\partial y} u_j + N_i \frac{\partial M_1}{\partial x} p_1 \right.$$

$$\left. - N_i \frac{1_{x1}}{F_R^2} + \frac{1}{R_E} \left(\frac{\partial N_i}{\partial x} \frac{\partial N_j}{\partial x} u_j + \frac{\partial N_i}{\partial y} \frac{\partial N_j}{\partial y} u_j \right) \right] dA^e \qquad (6.26)$$

$$- \int_{\angle_1^e} \frac{1}{R_E} N_i \frac{\partial N_j}{\partial n} u_j \, d\angle - \int_{\angle_2^e} \frac{1}{R_E} N_i \frac{\partial u_j}{\partial n} \, d\angle = 0$$

where $\angle_2^e$ denotes the boundary over which $\partial u / \partial n$ is specified, and $\angle_1^e + \angle_2^e = \angle^e$.

The momentum equation in the y direction is obtained by interchanging x, y and u, v. The continuity equation becomes

$$\sum_1^{n^e} \int_{A^e} M_i \left(\frac{\partial N_j}{\partial x} u_j + \frac{\partial N_j}{\partial y} v_j \right) dA^e = 0 \qquad (6.27)$$

Following an accepted practice [4, 5], the variation in pressure by shape functions one order lower than those for the velocity distribution gives

$$u = \sum_{i=1}^{i=n} N_i u_i \qquad (6.28)$$

$$v = \sum_{i=1}^{i=n} N_i v_i \qquad (6.29)$$

$$p = \sum_{i=1}^{i=m} M_i p_i \qquad \begin{array}{l} n = 8 \\ m = 4 \end{array} \qquad (6.30)$$

The same element geometry may be used where all eight nodes of a quadrilateral are associated with velocity and only the four corner nodes with pressure.

The assembled matrix equations take the form [2]

$$A\lambda = F + B \qquad (6.31)$$

where the chosen form for λ is

$$\lambda_i = \begin{Bmatrix} u_i \\ p_i \\ v_i \end{Bmatrix} \qquad (6.32)$$

Each coefficient in the matrix A has the form given in Ref. [2].

Natural boundary conditions are imposed in the F vector [2] as

$$f = \sum_1^{ne} \int_{A^e} \begin{Bmatrix} f_1 \\ f_2 \\ f_3 \end{Bmatrix} da \tag{6.33}$$

where

$$f_1 = N_i \frac{1_{x1}}{F_R^2}, \qquad f_2 = 0, \qquad f_3 = N_i \frac{1_{x2}}{F_R^2}$$

Similarly, the second component of the right-hand side is [2]

$$b_i = \sum_1^{ne} \int_{\angle_2^e} \begin{Bmatrix} b_1 \\ b_2 \\ b_3 \end{Bmatrix} d\angle \tag{6.34}$$

where

$$b_1 = \frac{1}{R_E} N_i \left[\left(\frac{\partial u_j}{\partial n} \right)^{\angle_e} d\angle \right], \qquad b_2 = 0$$

and

$$b_3 = \frac{1}{R_E} N_i \left[\left(\frac{\partial v_j}{\partial n} \right)^{\angle_e} d\angle \right]$$

ELEMENT TYPES

Elements for fluid flow are categorized by the velocity-pressure approximation allowed in the element. The finite element equations are obtained by variational or weighted residual methods. The most critical step in the analysis is the choice of adequate interpolation functions. The interpolation functions are characterized by the shape of the finite element and order of approximation. Various types of finite elements are described in Chapter 2. The quadrilateral element is used throughout this chapter.

The linear variation of a variable within an element is expressed by data provided at corner points. For quadratic variation, nodes are added midway between the corner nodes, as shown in Figure 6.1. Two-dimensional finite elements have been used widely because many domains can be idealized by two-dimensional elements.

Isoparametric Linear Element

The polynomial expansion for quadrilateral elements is often incomplete for the desired inclusion of side and interior nodes. The use of Lagrange or

Hermite functions allows the desired interpolation functions to be constructed simply. The term isoparametric is derived from the fact that the same parametric function that describes the geometry may be used for interpolating spatial variations of a variable within an element. The isoparametric element utilizes nondimensionalized coordinates. The four-noded rectangular element has a spatial variation of the variable p over the element defined by

$$p = \alpha_1 + \alpha_2 x + \alpha_3 y + \alpha_4 xy \tag{6.35}$$

or

$$\{p\} = [1 \; x \; y \; xy] \{\alpha\} \tag{6.36}$$

Each of the four unknown α values can be evaluated in terms of p at the node points. For the four nodal points, this variation is written as

$$\begin{Bmatrix} p_1 \\ p_2 \\ p_3 \\ p_4 \end{Bmatrix} = \begin{bmatrix} 1 & x_1 & y_1 & x_1 y_1 \\ 1 & x_2 & y_2 & x_2 y_2 \\ 1 & x_3 & y_3 & x_3 y_3 \\ 1 & x_4 & y_4 & x_4 y_4 \end{bmatrix} \begin{Bmatrix} \alpha_1 \\ \alpha_2 \\ \alpha_3 \\ \alpha_4 \end{Bmatrix} \tag{6.37}$$

or

$$\{P\} = [c] \{\alpha\} \tag{6.38}$$

To evaluate the α's we have

$$\{\alpha\} = [c]^{-1} \{p\} \tag{6.39}$$

Upon substituting for α we obtain

$$\{p\} = [1 \; x \; y \; xy] [c]^{-1} [p] = [N_1 \; N_2 \; N_3 \; N_4] \{p\} \tag{6.40}$$

$$N_1 = \frac{1}{4} (1 - \xi)(1 - \eta) \tag{6.41}$$

$$N_2 = \frac{1}{4} (1 + \xi)(1 - \eta) \tag{6.42}$$

$$N_3 = \frac{1}{4} (1 + \xi)(1 + \eta) \tag{6.43}$$

$$N_4 = \frac{1}{4} (1 - \xi)(1 + \eta) \tag{6.44}$$

or

$$N_i = \frac{1}{4}(1 + \xi_i\xi)(1 + \eta_i\eta) \; i = 1, \ldots, 4 \tag{6.45}$$

Isoparametric Quadratic Element

Consider the eight-noded rectangular element shown in Figure 6.1. The spatial variation of the variable u or v over the element is defined by

$$u = \alpha_1 + \alpha_2 x + \alpha_3 y + \alpha_4 xy + \alpha_5 x^2 + \alpha_6 y^2 + a_7 xy^2 + \alpha_8 x^2 y \tag{6.46}$$

where each of the eight unknowns α values can be evaluated in terms of u at the node points. For the eight nodal points, this variation is written as

$$\begin{Bmatrix} u_1 \\ u_2 \\ u_3 \\ u_4 \\ u_5 \\ u_6 \\ u_7 \\ u_8 \end{Bmatrix} = \begin{bmatrix} 1 & x_1 & y_1 & x_1y_1 & x_1^2 & y_1^2 & x_1y_1^2 & x_1^2y_1 \\ 1 & x_2 & y_2 & x_2y_2 & x_2^2 & y_2^2 & x_2y_2^2 & x_2^2y_2 \\ 1 & x_3 & y_3 & x_3y_3 & x_3^2 & y_3^2 & x_3y_3^2 & x_3^2y_3 \\ 1 & x_4 & y_4 & x_4y_4 & x_4^2 & y_4^2 & x_4y_4^2 & x_4^2y_4 \\ 1 & x_5 & y_5 & x_5y_5 & x_5^2 & y_5^2 & x_5y_5^2 & x_5^2y_5 \\ 1 & x_6 & y_6 & x_6y_6 & x_6^2 & y_6^2 & x_6y_6^2 & x_6^2y_6 \\ 1 & x_7 & y_7 & x_7y_7 & x_7^2 & y_7^2 & x_7y_7^2 & x_7^2y_7 \\ 1 & x_8 & y_8 & x_8y_8 & x_8^2 & y_8^2 & x_8y_8^2 & x_8^2y_8 \end{bmatrix} \begin{Bmatrix} \alpha_1 \\ \alpha_2 \\ \alpha_3 \\ \alpha_4 \\ \alpha_5 \\ \alpha_6 \\ \alpha_7 \\ \alpha_8 \end{Bmatrix}$$

$$\tag{6.47}$$

or

$$\{u\} = [c]\{\alpha\} \tag{6.48}$$

To evaluate the α's we have

$$\{\alpha\} = [c]^{-1}\{p\} \tag{6.49}$$

Upon substituting for α we obtain

$$\{u\} = [1 \; x \; y \; xy \; x^2 \; y^2 \; xy^2 \; x^2y] \, [c]^{-1} \, [u] \tag{6.50}$$

or

$$\{u\} = [N_1 \; N_2 \; N_3 \; N_4 \; N_5 \; N_6 \; N_7 \; N_8] \{p\} \tag{6.51}$$

The shape functions for the corner nodes are

$$N_i = \frac{1}{4}(1 + \xi_i - \xi)(1 + \xi_i\xi)(\xi_i\xi + \eta_i\eta - 1) \tag{6.52}$$

and the midside nodes are

$$N_i = \frac{1}{2} (1 - \xi^2) (1 + \eta_1 \eta), \qquad \xi_i = 0 \tag{6.53}$$

$$N_i = \frac{1}{2} (1 + \xi_i \xi) (1 - \eta^2), \qquad \eta_i = 0 \tag{6.54}$$

MATERIAL PROPERTIES

The basic variables in fluid mechanics are the three velocity components and two thermodynamic properties. Any two of the thermodynamic properties, such as pressure, temperature, and density, will determine the state and hence all other properties.

Newtonian Fluids

All fluids have viscosity which causes friction. Viscosity is a measure of the fluid's resistance to shear when the fluid is in motion. Viscosity μ is the proportionality between shear stress and velocity gradient. The ratio of viscosity to mass density ρ is called the kinematic viscosity γ. The viscosity of a liquid decreases with increasing temperature, whereas gases increase their viscosity with increasing temperature. This proportional relationship between shear stress and velocity gradient is known as the Newtonian relationship.

Non-Newtonian Fluids

Fluids that do not obey the linear relationship between shear stress and shear strain are generally named non-Newtonian fluids. Typical viscosity relationships for non-Newtonian fluids are shown in Figure 6.2.

Examples of non-Newtonian fluids are paints, polymers, food products (such as honey and applesauce), emulsions of water and oil, and suspensions of various solids and fibers used in well drilling.

Bingham plastics

Bingham plastics exhibit a yield stress at zero shear rate and then a straight-line relationship between shear stress and shear rate. The equation for a Bingham plastic is

$$\tau = \tau_y + \mu_p \dot{\gamma} \tag{6.55}$$

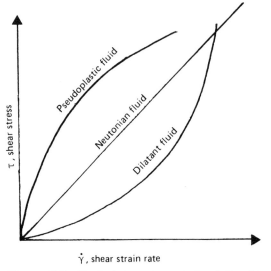

$\dot{\gamma}$, shear strain rate

Figure 6.2 Typical shear stress-strain relationships for non-Newtonian fluids.

where

τ = shear stress

τ_y = yield stress

μ_p = plastic viscosity

$\dot{\gamma}$ = shear strain rate

Pseudoplastic fluids

The pseudoplastic fluid has a decreasing slope of shear stress versus shear rate. One simple relationship used to describe pseudoplastic fluid is the power law. It may be written as

$$\tau = K\dot{\gamma}^n \qquad\qquad (6.56)$$

where

τ = shear stress

$n < 1.0$

K = measure of the consistency of the fluid

n = measure of how the fluid deviates from Newtonian fluid

$\dot{\gamma}$ = shear strain rate

Dilatant fluids

The dilatant fluid may also be represented by the power law where the exponent n is now greater than unity.

EXCITATIONS

A flow field can be created either by a stationary object with the fluid flowing about it or by an object moving through the gaseous liquid which causes the flow field. For a stationary object the fluid may flow around it owing to gravity forces or pressure. The pressure in a static fluid is defined as the normal compressive force per unit area acting on a surface. When the fluid moves, there may exist not only pressure forces in the fluid, but shear forces or stresses as well. For a moving object the flow field is caused by the difference of velocities.

BOUNDARY CONDITIONS

The implementation of boundary conditions into the resulting matrix will be considered next. The boundary conditions are: essential—where the variable is prescribed, and, natural—where the first-order gradient in the variable is prescribed.

Inviscid Potential Flow

Consider the two-dimensional implementation of boundary conditions for potential flow. The essential boundary condition is straightforward. Consider a matrix of the form

$$\begin{bmatrix} a_{11} & a_{12} & a_{13} & a_{14} \\ a_{21} & a_{22} & a_{23} & a_{24} \\ a_{31} & a_{32} & a_{33} & a_{34} \\ a_{41} & a_{42} & a_{43} & a_{44} \end{bmatrix} \begin{Bmatrix} \phi_1 \\ \phi_2 \\ \phi_3 \\ \phi_4 \end{Bmatrix} = \begin{bmatrix} F_1 \\ F_2 \\ F_3 \\ F_4 \end{bmatrix} \qquad (6.57)$$

If ϕ_3 is known, say $\phi_3 = K$, then this matrix can be rewritten as

$$\begin{bmatrix} a_{11} & a_{12} & 0 & a_{14} \\ a_{21} & a_{22} & 0 & a_{24} \\ 0 & 0 & K & 0 \\ a_{41} & a_{42} & 0 & a_{44} \end{bmatrix} \begin{Bmatrix} \phi_1 \\ \phi_2 \\ \phi_3 \\ \phi_4 \end{Bmatrix} = \begin{bmatrix} F_1 - a_{13}\,K \\ F_2 - a_{23}\,K \\ K \\ F_4 - a_{43}\,K \end{bmatrix} \qquad (6.58)$$

If the boundary condition $\phi_3 = 0.0$ this matrix becomes

$$\begin{bmatrix} a_{11} & a_{12} & a_{14} \\ a_{21} & a_{22} & a_{24} \\ a_{41} & a_{42} & a_{44} \end{bmatrix} \begin{Bmatrix} \phi_1 \\ \phi_3 \\ \phi_4 \end{Bmatrix} = \begin{bmatrix} F_1 \\ F_2 \\ F_3 \end{bmatrix} \tag{6.59}$$

The natural boundary condition is straightforward. The integrated value over the boundary associated with an element forms part of the equation. It appears as a contribution to the flux vector.

On the boundary $\angle$ where a flux is defined, the natural boundary condition can be written as

$$F_i = \int_{\angle} N_i \frac{\partial \phi_i}{\partial n} \, d\angle \tag{6.60}$$

The integrated value over the boundary associated with an element forms part of the overall equation and appears as a contribution to the flux vector.

Viscous Flow

When considering viscous flow using the velocity and pressure u, p, v formulation, there are two types of boundary conditions. The essential boundary condition defines u, p, and v on a surface. These boundary conditions are handled as described in the preceding section. The natural boundary condition defines the flux $\dfrac{\partial u}{\partial x}$ or $\dfrac{\partial y}{\partial v}$ on a surface or boundaries where natural boundary conditions are imposed. They appear in the right-hand vector as [2]

$$f_i = \int_A \begin{Bmatrix} f_1 \\ f_2 \\ f_3 \end{Bmatrix} \, dA \tag{6.61}$$

where

$$f_1 = N_i \frac{1_{x1}}{F_R^2} \tag{6.62}$$

$$f_2 = 0 \tag{6.63}$$

$$f_3 = N_i \frac{1_{x2}}{F_R^2} \tag{6.64}$$

Initial Conditions

For many problems the initial condition can be assumed to be zero. This is particularly true at a low Reynolds number. At a high Reynolds number you can save computer time by using a reasonable estimate for the initial values. It is also desirable to be able to stop the calculation after a specified number of iterations to ascertain whether the procedure is converging. Then the calculation can be restarted with the final iteration as the initial condition.

RESULTS AND VERIFICATION

The application of the Navier-Stokes equation to problems of interest to today's engineer involves the solution of highly nonlinear equations. The nonlinear nature of the equations dictates the use of iterative techniques to produce a solution. The results of these solutions to the Navier-Stokes equations must be examined throughout the solution process. Various convergence criteria are often set up to help identify the soundness of the solution.

EXAMPLES

Results from a flow analysis can include the velocity vectors, pressure contours, streamline vorticity, and temperature contour plots. Other results, such as tangential and normal stresses, heat fluxes, and flow rates at element boundaries, can also be determined.

The following examples were prepared using the FIDAP package, which is developed by Fluid Dynamics International. FIDAP is a general-purpose finite element program for simulating viscous incompressible flow, including the effects of heat transfer, in 2D, axisymmetric, and 3D geometries.

Flow Past a Car

The flow past a car from rest to 60 mph is modeled. Figure 6.3 shows the finite element mesh for the entire region and a closeup of the region immediately behind the car. Figure 6.4 shows the streamline contour plot for the entire model and the velocity vector solution for the closeup region. Both plots are at a time of 2.57 sec.

Secondary Flow in Rotating Cylinder

The secondary flow is modeled in a rotating cylinder. The rotational flow component has been removed. The cylinder is rotating and nutating. Figure 6.5a shows the superposition of velocity vectors plus speed contours in a plane perpendicular to the longitudinal area. Figure 6.5b shows the speed contour plot on a longitudinal plane.

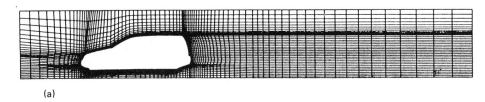

(a)

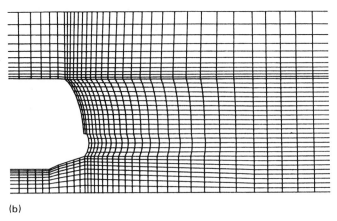

(b)

Figure 6.3 (a) Finite element mesh plot of car and surrounding air. (b) Closeup of the region behind the car.

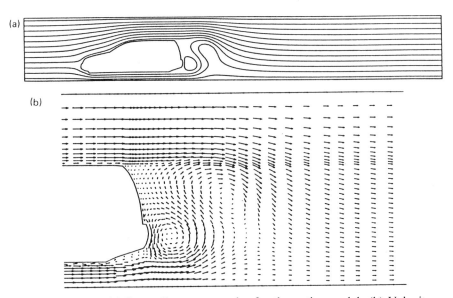

(a)

(b)

Figure 6.4 (a) Streamline contour plot for the entire model. (b) Velocity vector plot for the closeup region.

(a)　　　　　　　　　　　(b)

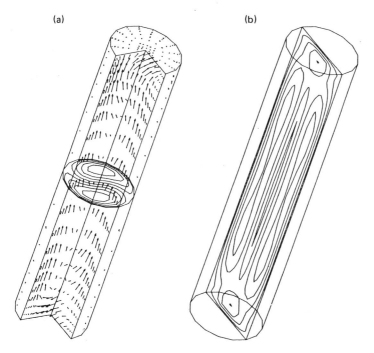

Figure 6.5 (a) Velocity vectors and speed contours. (b) Speed contours.

Flow Past a Step in a Channel

A heated cube in a channel with air being forced by from back to front is analyzed. Figure 6.6a shows the mesh with the corner removed. The mesh is shown using hidden line-plotting techniques. This is typical of what is needed for three-dimensional finite element models to make sense. Analysis results (Figure 6.6b) show the velocity vector solution.

Pasteurization Process

An axisymmetric transient analysis of the in-bottle pasteurization of beer is performed. The temperature at the outside of the fluid is a function of time the analysis took at the temperature history of the tunnel pasteurizer.

The pasteurizer sprays progressively hotter water in the beer bottles as they pass by on the moving belts. Figure 6.7 shows the solution plots of the velocity vectors and temperature contours at various time steps through the transient.

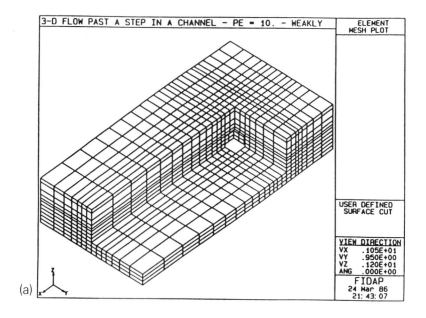

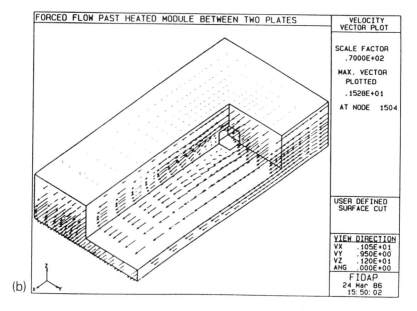

Figure 6.6 (a) Finite element mesh plot. (b) Velocity vectors.

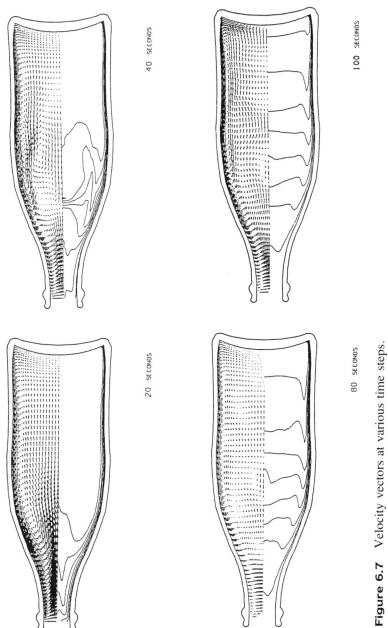

Figure 6.7 Velocity vectors at various time steps.

REFERENCES

1. Steele, J. M., "Manager's Guide to Finite Element Analysis," *Computer-Aided Engineering,* June 1986.
2. Taylor, C., and T. G. Hughes, *The Navier-Stokes Equation.* Pineridge Press Limited, Swansea, England.
3. Baker, A. J., *Finite Element Computational Fluid Mechanics,* McGraw-Hill Book Co., Hemisphere Publishing Corp., New York.
4. Gresho, P. M., and R. L. Lee, "Don't Suppress the Wriggles—They're Telling Us Something," in Hughes, T. J. R. (Ed.), *Finite Element Methods in Convection Dominated Flows,* Winter Annual Meeting A.S.M.E., 1979.
5. Hood, P., and C. Taylor, "Navier Stokes Equations Using Mixed Interpolation," Proc. 1st Int. Conf. on F.E.M. in Flow Problems, Swansea, England, 1974.
6. Gallagher, R. H., *Finite Elements in Fluids,* Vol. 1 and 2, John Wiley & Sons, Inc., New York, 1975.
7. Gallagher, R. H., *Finite Elements in Fluids,* Vol. 5, John Wiley & Sons, Inc., New York, 1984.
8. Gallagher, R. H., *Finite Elements in Fluids,* Vol. 1—*Viscous Flow and Hydrodynamics,* John Wiley & Sons, Inc., New York.
9. Girault, and P. A. Raviart, *Finite Element Approximation of the Navier-Stokes Equations,* Springer-Verlag, New York, 1979.
10. Thomasset, F., *Implementation of Finite Element Method for Navier-Stokes,* Springer-Verlag, New York, 1981.
11. Schlichting, H., *Boundary Layer Theory,* Pergamon Press, Elmsford, NY, 1955.

7 Hardware-Software Considerations

Glenn H. Stalker

The MacNeal-Schwendler Corporation, Milwaukee, Wisconsin

HARDWARE CONSIDERATIONS

As is evidenced by the preceding chapters, finite element analysis is applicable to many branches of engineering. It represents one of the most powerful tools available to the engineer for the solution of engineering problems.

Since it is the engineer who bears the responsibility for the validity of the results of a particular analysis, it should be the engineer who decides which tool is appropriate. In finite element analysis, these tools take the form of both hardware and software. Hardware available to the engineer spans the entire range from hand-held calculators to supercomputers. This has not always been the case. Over the past 30–40 years, the technical revolution in the electronics circuitry field has provided increasingly attractive price-to-performance ratios for computer hardware. Consequently, computer hardware and software applicable to finite element analysis have developed to fit a wide range of needs and budgets.

History

The first digital computers arrived on the scene in the 1940s. These were truly "mainframes" in that they were huge electromechanical devices functioning with clicking relays. These early machines could multiply two 10-digit numbers at the astounding rate of 6 sec per operation. When vacuum tubes replaced the relays and mechanical moving parts, the speed for an operation was increased to about 1/40 of a second, and to 1/2000 of a second by the mid-1950s [1]. The vacuum tube computers of the 1950s had computational abilities that were only slightly better than those of the programmable hand-held calculators of today. [5]

The transistors that began replacing the vacuum tubes in the late 1950s were more compact and produced only a fraction of the heat. They also had the added advantage of being cheaper and far more reliable. The speed of these machines was now in the range of 100,000 instructions per second.

With the integrated circuitry of the 1970s, a single silicon chip could contain as much horsepower as an entire mainframe computer 20 years earlier. Computers could now perform operations at the rate of millions of instructions per second. That trend has continued into the 1980s, bringing incredible computing power to individual workstations and even personal computers. As technology continues to leapfrog in computer chip design, it is all the more imperative that the engineer be aware of the wide variety of both hardware and software tools available.

State-of-the-Art

Prior to the late 1970s, most finite element work was performed on mainframe computers. Since then, an increasing number of finite element codes have begun migrating to engineering workstations and personal computers. As one might expect, the capabilities and problem sizes are somewhat limited for the smaller computers. This is not to say that these are inferior tools, but that the wide variety of engineering tasks can now be solved by a wider range of alternative methods.

For purposes of this discussion, all computers will be divided into three general categories: mainframes, workstations, and personal computers. Within the category of mainframes will be the supercomputers (CRAY X-MP, CRAY-1S, IBM 3090), the minisupercomputers (CONVEX C-1, FPS-164/364, ALLIANT FX/8), and the newer breed of mainframes (IBM 3033 and 3081, DEC VAX 11/780, 11/785, 8600, 8650, and 8800). The workstation category includes the superminis (Apollo DN660, VAX 11/750, HP 9000) and the minis (Apollo DN3000, MICRO VAX II). The personal computers are primarily IBM and its clones (IBM RT, AT, XT, and compatibles). Manufacturers' names and models are mentioned here not to provide an all-inclusive list, but to give the reader a frame of reference for the variety of engineering-related computer hardware. In each of these categories there is extensive overlap between the high end of one segment and the low end of the next.

As each segment has experienced increases in speed, it has spawned a lower-level machine for which finite element analysis is adaptable. The increased capabilities of mainframes left room for the workstation market. Workstation enhancements have, in turn, provided the niche for personal computers.

Personal computers

Personal computers (PC) or microcomputers have begun to capture a large market share for simple types of finite element analysis. Activities that have come down to the PC are CAD-related software to build coarse models, run small problems, and interpret results. These problems previously would have been hand-calculated by means of a "closed-form" solution, idealizing the problem to fit textbook equations. With the finite element software available on hardware at the engineer's fingertips, he is much more inclined to input the more realistic geometry and loading to obtain better rough-cut results. A detailed finite element analysis might still be necessary, but now it takes the form of refinement or weight reduction rather than major design evaluation.

A big advantage to FEA software on the PC is that it is low cost and ideally tailored to the infrequent user. This implies that the commands are more like English and require less hunting through a users' manual when attempting an analysis once every 3 months. Typically, the software provided here is most appropriate in the early design phase, when various different design schemes might need to be evaluated.

As PCs continue to increase in speed and their graphic displays improve in resolution, they will grow in usage in model building and results evaluation. The key to these two tasks is response time. Once the user response time matches that currently attainable on the engineering workstation, these functions will certainly be downloaded to the PC.

Today's PC hardware is appropriate both for those companies whose FEA needs are so infrequent and casual that they don't warrant internal staff or sophisticated hardware, and for those larger companies that might utilize it in the early design evaluation phase. Future higher-speed PC hardware will be appropriate for much of the FEA work processed today on workstations.

Workstations

All FEA functions that previously were reserved exclusively for the mainframe have been brought down to the engineering workstation. Even though this might have been at the expense of some speed, the workstation generally provides more user control. The mainframe has historically been a multiuser, remote system, controlled by someone else, whereas the engineering workstation tends to be a single-threaded, one-user system, at the same site as and under the control of the engineer. These factors along with the increased speed and capabilities of the minicomputers have fueled their attractiveness.

The basic engineering workstation includes the central processing unit (CPU), the monitor (color or monochrome), and one or more formats of data entry (keyboard, digitizing tablet, light pen, or mouse). The station must either contain its own disk drive or have access to one if it is part of a larger network. While the speed of the CPU may be one of the initial concerns with this hardware, inadequate disk space will almost surely follow. Finite element analysis of ever-increasing sizes of models taxes both the CPU for acceptable run times as well as disk space to store the massive amounts of data calculated. A cost/benefit analysis must occasionally be conducted to determine whether the system warrants expenditure for additional CPU memory and/or disk storage capacity.

Most workstation monitors sold today for use in a finite element environment are color. The benefits derived from differentiation of model regions, properties, element types, and the like in preparing a model, and from results evaluation, easily justify the incremental cost of color over monochrome. Monitors are typically of the raster type, in which an image is created on a matrix of tiny dots called pixels. An electron beam scans the entire screen, illuminating selected pixels in an on/off pattern to display the desired image stored in the computer's memory. The one drawback to the color raster screen is that diagonal straight lines appear stair-stepped or somewhat jagged because of the discrete nature of the pixel matrix. This was not a problem with the older-technology, higher-resolution storage tubes, which would trace lines with beams of electrons, illuminating the phosphorous screen indefinitely. The disadvantages of storage tubes are that they are, first, monochrome and, second, do not retrace the image constantly, preventing the animation possible with a raster screen. A third type of monitor that is in use is the vector refresh. This borrows qualities from both the raster and the storage tubes in that it constantly repaints and it traces lines from end to end. Because it constantly repaints, it can animate. Also, because it traces from point to point, it has much better resolution than the raster. Its primary disadvantages are flicker with large models and lack of color. Despite their resolution deficiency, color raster monitors tend to be the choice for most FEA workstations.

Communication with the computer is accomplished through data entry on a keyboard. Aside from a keyboard, which is a standard item, many systems include either a digitizing tablet, a light pen, or a mouse for user convenience of picking standard menu entries or items on the screen. The determination of which of these three will be used is generally software dependent.

Mainframes

This segment of hardware probably has the widest range of speeds and resulting problem execution times. For purposes here, it will be defined as those machines that are faster than the workstation and, unlike the workstation, are utilized by more than one person at a time.

Mainframes generally have more CPU power and more disk space than their workstation counterparts, but also have more users vying for those scarce resources. So, to a certain extent the problems of time to completion and disk space are a common concern.

The most efficient use of mainframes in the finite element field is to solve large problems quickly. They are best suited to solving one or more batch jobs at a time, quickly, and are least appropriate for interactive use. It can be advantageous to do pre- and postprocessing on a work station and move the data to a mainframe just for the solution phase. This becomes even more the case as the speed of the mainframe approaches that of the CRAY.

Future

The trend for the future of finite element analysis-related hardware is an exciting scenario for the engineer. Gone will be the days when designs are approximated by crudely representative closed-form solutions. Gone will be the days when FEA is too sophisticated for the novice user. And gone will be the days when FEA is out of reach financially for anyone other than the largest companies. For the most part, these trends are already well underway to becoming a reality today.

The batch solution of problems is today available on most hardware. Larger machines are better suited to the huge, more complicated analyses, while the smaller machines can handle the crude, rough-cut, simplistic studies. The more powerful the smaller computers become, the more they will be utilized in larger, sophisticated analyses. The interactive portion of FEA is becoming increasingly applicable to minicomputers and the PC. In the not too distant future most pre- and postprocessing will migrate to the PC-sized computer.

SOFTWARE CONSIDERATIONS

When considering what sort of FEA software is appropriate, three general categories need to be investigated: (1) preprocessors, (2) model solution, and (3) postprocessors. Preprocessors interactively assist the engineer to formulate the geometry, the material properties, the boundary conditions, and the loading to be imposed. Once the input data deck has been assembled with the assistance of the preprocessor, the solution of simultaneous equations is provided by the FEA program's solver. Since stacks of pages of printed output are generally difficult to interpret, a postprocessing program is utilized to reformulate the output into more understandable terms, such as color plots, deflected shapes, or graphs.

History

As was mentioned in Chapter 1, the term "finite element" was first used in a paper on plane elasticity problems by Clough in 1960. The solution of plane stress problems by the method of dividing the structure into triangular regions and solving the equations of elasticity was presented in a paper by Turner, Clough, Martin, and Topp in 1956. This method was coined the "direct stiffness method." [4] While these were the first evidences of the terms of the FEA method in the literature, the actual idea had its inception somewhat earlier. In the 1940s, the mathematician Courant, along with physicists and engineers, independently proposed solution to these problems by means of "piecewise continuous functions defined over triangular domains." [4]

The FEA method is a systematic technique that found its practical realization in computer programs. Were it not for the advent of the digital computer, mathematicians and engineers would doubtless have continued the evolution of traditional methods of analysis. For instance, in the analysis of frames and beams the following methods were developed: Moment of Area (1874), Strain Energy (1879), Slope Deflection (1892), and Moment Distribution (1930) [2]. Each method sought to address the shortcomings of the previous one, whether those were restrictions in their application or difficulty with the voluminous calculations required. Most subsequent methods were born out of a need to simplify the calculations of traditional engineering problems rather than refinement to the theory on which they were based. Once the computer arrived on the scene, emphasis once again shifted toward obtaining a more correct solution, and away from minimizing calculation effort.

In its infancy in the early 1970s, most of the commercially available finite element software resided on mainframes. The majority of pre- and postprocessing was done by hand. Coding forms of input data was transformed into punched cards which served as the input data deck to the FEA solver. The extent of preprocessing programs were batch programs designed to plot model geometry as a check of the input data. Output from the FEA program consisted of pages and pages of printed output, with special-purpose programs to sort some of these data. Postprocessing might have included batch procedures for plotting deformed shapes of the model. Interactive pre- and postprocessing was generally not available until the late 1970s.

State of the Art

Some of the key programs available today are listed in Table 7.1. The last 10–15 years have seen dramatic improvements in user-friendliness, computational capabilities, and ease of results evaluation. Provided the geometry is fairly simple, it is possible to describe the model geometry, develop an acceptable mesh, submit the model for solution, and interrogate the results of an analysis, in less than a day. Unfortunately, many of the geometric shapes that require FEA can only be described with days or weeks of effort. If an analysis is to take an extended period of time to complete, it is generally the model generation phase that is extensive.

Preprocessors

Preprocessors have as their primary function to prepare the finite element model for solution. The most time-consuming aspect of this process is the definition of the geometry and description of the mesh on that geometry. While input of the material properties, constraints, and loads are essential, they do not require as much time.

A good finite element preprocessor will have many of the features available

Table 7.1 Some Representative Finite Element Software Packages

Name	Origin	Type (primary)
ABAQUS	H.K.S.	Structural, especially nonlinear
ANSYS	Swanson Analysis Systems, Inc.	Structural
STRUDL	M.I.T.	Structural, civil
McAUTO/STRUDL	McDonnell-Douglas Automation	Structural, civil
NASTRAN	N.A.S.A.	Structural
MSC/NASTRAN	MacNeal-Schwendler Corp.	Structural
STARDYNE	Systems Dev. Corp.	Structural
FIDAP	Fluid Dynamics	Fluid
MSC/MAGNETIC	MacNeal-Schwendler Corp.	Electromagnetic
PDA/PATRAN	P.D.A.	Pre- and postprocessor
SUPERTAB	S.D.R.C.	Pre- and postprocessor
MARC	MARC Software Intern., Inc.	Structural

on a CAD system. These are essential, particularly when complex geometries need to be represented. If the necessary CAD features are not present, there needs to be a translator available, through which CAD wire frame data could be transmitted. If only relatively simple geometries are to be modeled, this is not of great concern.

Once a geometric wire frame of the object has been represented, the surfaces and/or volumes of the structure are laid out. The mesh for the model is then mapped onto these surfaces/volumes. Different properties of the model are described as the meshing takes place. Loads and boundary conditions can be applied and graphically verified.

Most preprocessors include various model-checking schemes to verify that the model has no missing elements, duplicate elements, severely warped elements, and so forth. With a color monitor, various portions of the model can be plotted by property ID to verify that all properties have been assigned correctly. Although most FEA programs themselves include a resequencing step to provide the most efficient numbering scheme for the model, it is sometimes helpful to have this as part of the preprocessor.

Solution modules

Whether the problem to be solved involves structural, thermal, electromagnetic, or fluid analysis, the solution procedures are quite similar. There is generally an input phase, an assembly and solution phase, and an output phase. Although it is not the intent of this book to provide a roadmap to construct a finite element code, it is helpful to review the major tasks of the code, so that the engineer can more effectively provide input to the program and interpret output from the program.

In the input phase, the program reads the input data, opens the necessary files to store data on the computer, and sorts the input data according to nodal coordinates, connectivity, material properties, loads, and constraints. Also in this phase is the assembly of the stiffness matrix based on nodal connectivity and material properties. Essential diagnostics are provided as part of this process: number of nodes, number of elements, and so forth, essentially echoing to the user what it is the computer understands to be the parameters of the problem. If any of these parameters are different from what the engineer intended, it is possible that the input data were somehow in error. It is also in this phase that errors in format of input are detected.

In the assembly and solution phase, the nodal equations that will provide solution to the unknown nodal values are assembled. The next step involves inversion of the stifness matrix to solve these equations by Gaussian elimination [4] or some other appropriate method of solving linear simultaneous equations. It is this phase of the program that generally consumes the most computer time and memory.

The output phase simply provides the user with a printout of the nodal unknowns (such as displacements for structural analysis) and the resulting element values (such as stresses for structural analysis). Also provided with this output is generally some sort of equilibrium or out-of-balance check which assesses the overall accuracy of the solution and identifies any possible ill-conditioning of the matrices (see pp. 64, 99, and 135). Finally, the output for graphic display is created. This generally takes the form of a binary file that can be read into the appropriate postprocessing software package. This file contains a duplicate of whatever results are available as printed output, in a format that can more easily be displayed graphically.

Postprocessors

The first rule in finite element analysis should be that the user should never have unquestioning trust in the results of a particular analysis. Instead, those results should be compared against hand calculations, engineering judgment, and common sense. A brief look at a few significant pieces of output can shed light on the validity of the solution. As mentioned previously, the overall accuracy and conditioning of the solution can be determined from the equilibrium or out-of-balance check. As this relative error term increases in value, less confidence should be put in the results.

Other quick checks that can be performed on the output are: scanning the displacements (nodal results) for any unreasonable values, plotting the deflected shape graphically, or plotting the stresses (element results) graphically. Any results that appear out of the ordinary or cannot be explained with the help of good engineering judgment or common sense should be further investigated. Postprocessing packages available today have made it almost too easy to jump to the final result without validating the accuracy of the solution first. The engineer must discipline himself to take the time to make the initial checks before jumping to the bottom line.

Commercially available postprocessing codes provide outstanding capabilities to graphically represent pertinent results. From static deformed shapes to animated displays, from color stress contour plots to movies of increasing stress patterns, the graphics of postprocessing have provided insight that numbers alone could never provide.

Future

The trend for improvement in the future of finite element analysis software almost certainly must be in the area of preprocessing. This is the segment of the project that tends to be the most time-consuming, and consequently the biggest deterrent to its use. More automated model-building routines to develop meshes from less specifically defined geometry are essential. Some form of artificial intelligence may be appropriate to duplicate some of the more tedious activities required to build finite element models.

REFERENCES

1. Krouse, John K., *What Every Engineer Should Know About CAD/CAM*, Marcel Dekker, Inc., New York, 1982.
2. Litton, E., *Automatic Computational Techniques in Civil and Structural Engineering*, Halsted Press, New York, 1973.
3. Livesley, R. K., *Finite Elements: An Introduction for Engineers*, Cambridge University Press, Cambridge, England, 1983.
4. Huebner, Kenneth H., *The Finite Element Method for Engineers*, John Wiley & Sons, Inc., New York, 1975.
5. Beakley, George C., and Robert E. Lovell, *Computation, Calculators and Computers*, Macmillan Publishing Co., Inc., New York, 1983.

8 Professional Practice and Future Potential

G. E. Barron

The MacNeal-Schwendler Corporation, Milwaukee, Wisconsin

The development of the finite element method of analysis has come a long way over the last thirty years. Its application is widespread in structural analysis and is increasing in other areas. The preceding chapters of this book have introduced finite element analysis and its history, documented its mathematical formulation for several areas of engineering endeavor, and provided real world examples solved using the technique. This chapter seeks to answer the 'What now?' part of what every engineer should know about finite elements. Two aspects of this will be examined briefly. One is the human side, the other some speculation about what improvements might be expected to push beyond the current state-of-the-art.

PROFESSIONAL PRACTICE

The human side of Finite Element Analysis (FEA) relates to the individual engineer using the technique, the management of the firm for which the analysis is performed, and the professional organizations which may influence the practice of FEA. When the software for using FEA was embryonic, say in the mid 1960s, and computers were small and slow, not many people were able to solve real problems and obtain valid answers. Those who did were rarely believed by those who built and tested, so little direct economic loss came from inaccuracies in the analysis. Further, those people who were involved in advocating finite elements were keenly aware of the need for diligence if any success was to be achieved. Model sizes tended to be small and the person who created the geometrical grid and element meshes was likely the same person who ran the analysis, reviewed the output, correlated analytical data to test and wrote the report. Under such circumstances, the integrity of the overall procedure was well maintained. Graphical displays were limited in those days to perhaps some monochrome plotter output and a complete paper printout of all input and output data, which was kept to back up the analysis report. Anyone could go back to check geometrical dimensions or verify which material properties were really used.

Today's situation is different in many ways. The brief discussion below is intended to illustrate the Professional Practice side of FEA and the pitfalls which are possible. In certain ways, the management of the FEA process has become as important as the actual tool itself, especially since the technique is now widely used in a production environment. Today model sizes are becoming quite large, probably an order of magnitude larger on average than fifteen to twenty years ago. Gone are the days when the model builder could memorize all the grid and element numbers during the course of an analysis and virtually become friends with each element. No longer are results interpreted by poring over printed output and looking at the stress value for element 101 and saying to oneself, 'Ah hah, the stress near the top of the pressure vessel just to the right of the little opening is well below yield'.

The situation today, in a larger firm, is more likely that a 10,000 degree-of-freedom model is built by one person and analyzed by another and that there will be several files on the computer containing minor geometrical and mesh changes (all automatically renumbered so no new model contains any direct correlation to the original). If analysis runs are made using a modern FEA software product with the printout data echo turned off,* then part of the audit trail is lost. If furthermore, some lethargy or forgetfullness sets in and the title card is not changed for two or three runs, then it is very possible that results will be obtained and reported for the wrong version of the model. Color graphics output from today's software is impressive, but only useful if a careful review has been made of the tabular output to validate the solution.

Good management policy and work practice to produce documentation with an audit trail to validate analysis results is very important in today's FEA environment. Computer data sets containing the models and corresponding output need to be archived and the data set names referenced in the reports.

---

*An analyst is motivated to do this because the large printout may be a meter high and no one wants to look through it or store it.

Similarly the drawing numbers from which the model was created need to be documented in the report. Did the original geometry for the model come from a Computer Aided Design system via some computer file transfer? If so, complete documentation of the equivalent drawing number and any revision number needs to be captured for inclusion in the FEA report.

Close communication is required between the engineer performing the FEA work and the designer of the device to be analyzed. If the design and its drawing change during analysis, then the designer must notify the FEA analyst. Also, the FEA analyst must spend sufficient time with the designer determining actual real world loads and material properties to be used for input to the FEA software product which is being used. Communication must also flow both ways in determining at the outset of the project what purposes FEA is to serve in validating the design.

Once the analysis is finished, the engineer who has done the FEA work must help the designer or design engineer responsible for the part or device understand the implication of the analysis results for the function of the design. If this communication is handled well, then the two should collectively arrive at insight neither might have obtained alone. At this point, mutually agreed changes can be rerun through the FEA software and changes in the performance determined. Typically, this is when the payback of having used the FEA process occurs. If the need for the expense and time delay of additional prototypes can be avoided and a better part or device sent into production, then FEA and the people who use it are performing a real service.

The point of the above discussion is that people and people management are key to producing good results in FEA applications. Emphasis has changed from the time when engineers asked 'Can FEA solve this problem at all?' and 'Are the results correct?' to whether FEA can become a viable part of the overall design and manufacturing process.

Today there are good FEA software products available commercially and one can pick from amongst them. The software vendors are continuing to test their software and publish application manuals which show that their FEA software correlates to both classical solutions and experimental results. This proof testing that the FEA software is accurate should continue. However, once an engineering firm or department selects the best tool they can obtain, the emphasis shifts to professional practice issues or people issues.

A 'What' worth knowing about professional practice is that at least two organizations in the world are concerned about quality results from FEA and the practice of FEA. They are:

1.   Project for Quality Aspects of FEA, started in 1984 by fifteen Dutch FE users community members. Results of this study are expected in 1987.

Contact: Secretary, CIAD, Kabe-project, t.a.r.ir. C. Tjallema, P. O. Box 74, 2700 Ab Zoefermeer, The Netherlands.

2. The National Agency for Finite Element Methods and Standards, started in the UK in 1982. Contact: W. M. Mair, Chairman NAFEMS, Dept. of Trade and Industry, National Engineering Laboratory, East Kilbridge, Glasgow, G75 OQU, Telephone 035 52 20222, Telex 777888.

It is hoped that this discussion will inspire readers affiliated with a professional engineers association or an educational institution to explore certification of FEA software products or FEA users. If any reader wants to explore ways in which an effort in the United States might be launched to perform functions similar to those listed above, feel free to contact the editor of this book in care of the publisher.

FUTURE POTENTIAL

Thoughts about where the future of the finite element method of analysis will go from here are offered by the authors of the previous chapters. The five authors of this book possess a collective seventy years of experience in Finite Elements after having first acquired solid engineering education in Aerospace, Civil, Electrical, and Mechanical Engineering. Thus the perspective is multidisciplinary. Imagine that this book has been a seminar on what every engineer should know about finite element analysis and that you are listening to concluding thoughts from a panel discussion.

G. E. Barron:

The productivity of performing FEA will improve dramatically when companies make the transition to full three dimensional (3D) computer aided design (CAD), rather than just producing two dimensional drawings. In another ten years we will likely see a workable integration of design, analysis, and manufacturing in most major companies. The automatic meshing of finite elements into the envelope specified by the 3D CAD system will take a quantum leap when artificial intelligence (AI) systems are integrated into the now totally mathematical meshing algorithms. There is a distinctive difference in the productivity of people who build models today using such tools as PATRAN, so it seems reasonable to add expert knowledge on how to transition elements, and so forth, to the software.

The trend toward economical super computers seems destined to continue, so non-linear analysis will become as common as linear analysis today. Two challenges will emerge from this:

1. Validation of a correct answer will be more difficult since non-linear problems may have more than one solution. Path dependent non-linear solutions will require characterization of the path, not just the final solution, to gain insight into the device performance.

2. Substantially more information on material properties than exists today will be needed to use the non-linear FEA capabilities.

Within ten years, FEA systems will become somewhat hybrid and will include the best features of other analysis techniques for the special cases where they are better. This will include the boundary element method, method of moments (for electromagnetics) and finite differences (for compressibility effects in flow or plasticity).

The finite element approach will continue to be the bedrock of engineering analysis because of its complete generality, especially in handling boundary conditions correctly, and not having to make limiting assumptions at the formulation stage. The aggravation of building large models will go away in time.

G. H. Stalker:

The future for hardware and software related to finite element analysis is indeed encouraging to the engineer. Hardware is continually becoming more powerful so that larger and more sophisticated analyses can be performed in increasing numbers. Where once there was only the pocket calculator and the mainframe computer, now there is an emerging multitude of hardware options to suit many needs and many budgets.

In FEA software, preprocessing continues to be the area that must improve for the future. Postprocessing has progressed markedly with color and animation. Solution modules are limited only by the computer hardware budget. Preprocessing improvements must be made in the areas of integration with drafting data and automated meshing of complex three dimensional (solid) shapes.

Dr. N. J. Lambert:

Fluid analysis is the newest area of FEA. It has the opportunity for greater widespread application as it becomes easier to use and the engineer becomes educated about it.

Future efforts will include enhancements to the computational techniques available for high Reynolds number flow. Much effort is required to provide stable solutions that converge quickly.

Dr. J. R. Brauer:

Electromagnetic finite elements will be used more extensively at higher frequencies, including optical frequencies. High frequency electromagnetic waves created by antennas and other sources will be analyzed.

Coupling between electromagnetic fields and mechanical motion will be analyzed, including the eddy currents created by the motion. Such analysis will be especially helpful for fast-acting solenoids.

Instead of requiring current as input, voltage sources will also be allowed. The series resistance of the voltage source would be included in the analysis.

Dr. V. D. Overbye:

The history of finite element analysis chronicled in Chapter 1 shows that structural analysis (Chapter 3) first occurred during the 1960s with widespread growth during the 1970s. Thermal analysis (Chapter 4) and electromagnetic analysis (Chapter 5) became increasingly popular during the 1980s. Fluid flow analysis (Chapter 6) is now becoming widespread and will see explosive growth during the 1990s. Turbulent flow analysis is still in its infancy and requires considerable attention on the part of finite element program developers to make the code user friendly.

No serious threat to the finite element method appears likely in the near future. The principles described in this book can be expected to serve the practicing engineer and student for many decades in the future.

Index

about the editor

JOHN R. BRAUER is Senior Consulting Engineer with The MacNeal-Schwendler Corporation, Milwaukee, Wisconsin. Prior to this position, he was a consulting engineer with the A.O. Smith Corporation, Milwaukee, Wisconsin, and a research assistant in the department of electrical engineering at the University of Wisconsin-Madison. The author of 40 technical articles, Dr. Brauer is a senior member of the Institute of Electrical and Electronics Engineers (IEEE), Chairman and Secretary of the IEEE Magnetics Society Milwaukee Chapter, a member of IEEE's IAS Electric Machines Committee, Tau Beta Pi, Eta Kappu Nu, and Sigma Xi. Dr. Brauer received the B.E.E. (1965) degree from Marquette University, Milwaukee, Wisconsin, and M.S.E.E. (1966) and Ph.D. (1969) degrees from the University of Wisconsin-Madison.